ENGINEERING TOOLS, TECHNIQUES AND TABLES

CALIBRATION TECHNOLOGY, THEORIES AND APPLICATIONS

ENGINEERING TOOLS, TECHNIQUES AND TABLES

Additional books in this series can be found on Nova's website under the Series tab.

Additional e-books in this series can be found on Nova's website under the e-books tab.

ENGINEERING TOOLS, TECHNIQUES AND TABLES

CALIBRATION TECHNOLOGY, THEORIES AND APPLICATIONS

IKUMATSU FUJIMOTO
AND
KUNITOSHI NISHIMURA
EDITORS

nova publishers
New York

For permission to use material from this book please contact us:
Telephone 631-231-7269; Fax 631-231-8175
Web Site: http://www.novapublishers.com

Additional color graphics may be available in the e-book version of this book.

Library of Congress Cataloging-in-Publication Data

ISBN: 978-1-62618-808-2
Calibration technology, theories and applications / editors, Ikumatsu Fujimoto and Kunitoshi Nishimura.
pages cm
Includes bibliographical references and index.
1. Calibration. I. Fujimoto, Ikumatsu, editor of compilation. II. Nishimura, Kunitoshi, editor of compilation.
QD75.4.C34C35 2013
621.37'2--dc23

2013016530

Published by Nova Science Publishers, Inc. † New York

CONTENTS

PREFACE

This volume aims to introduce the technique, theory and application of the calibration in various fields.

The term "calibration" is strictly defined in the international vocabulary of metrology (VIM) published by Joint Committee for Guide in Metrology (JCGM) organized by eight international organizations, BIPM, IEC, IFCC, ISO, IUPAC, IUPAP, OIML, and ILAC. Their definition is as follows:

"The calibration is an operation that, under specified conditions, in a first step, establishes a relation between the quantity values with measurement uncertainties provided by measurement standards and corresponding indications with associated measurement uncertainties and, in a second step, uses this information to establish a relation for obtaining a measurement result from an indication."

In simple words, the calibration is generally an operation to identify the state of an instrument (e.g. its accuracy, function and performance) and to determine a relationship between an output (measuring value) and an input of the instrument, or an essential value of a measuring object under specified conditions in a measuring system. Furthermore, it can be also regarded as a comparison of measuring equipment against a standard instrument of higher accuracy.

In the measuring field, the calibration is clearly confirmed in the traceability system. The traceability is defined as the property of a result of a measurement whereby it can be related to strictly defined standards, in general nationally or internationally recognized, through an unbroken chain of comparisons. Namely, all calibrations are strictly defined to be performed traceable to a nationally or internationally recognized standard. However, except for in the measuring field, the concept of traceability in not always defined.

Even though a calibration can be essentially different from an operation for adjustment and/or compensation, the term "calibration" is frequently used to describe an operation to adjust and/or compensate a measuring system. It should be noted once more that this does not apply to the measuring research institutions adhering to the strict definition of traceability.

Furthermore, calibration sometimes has a broader meaning, and can refer to the method for accurately determining the relational expression between an input/output of an instrument. In this case, a research area pursuing such an easy calibration method in which the lower cost specimen is used instead of highly accurate standards has also become active. This calibration method is called autonomous calibration or self-calibration.

Finally, global optimization methods and Monte Carlo simulations are also utilized to determine accurately the outcomes of complicated and large-scaled calibration models, because current computers can process massive amount of data in a short period.

The evaluation whether the calibration result is acceptable or not is conclusively carried out by uncertainty analysis through a series of iterative operation for adjustment and/or compensation of the difference between measuring value and calibration result. Namely, standard uncertainties of input variables are identified in a first step and a combined standard uncertainty is computed according to the formula of error propagation in a next step. The uncertainty is finally expressed as the expanded uncertainty. This process is general and described in Chapter 1-2, and 4.

The traceability system seems to be not established in the fields of meteorology and economic. A number of complicated metrogical and economic models are presented in Chapter 5-6, and 9. Since the meteorological and economic "world" is too complicate and has too many parameters, a precise modeling is very difficult.

The definition of the calibration in this volume refers to the above-mentioned broad sense of the term. Several recent topics applying the definition of calibration in their respective fields (metrology, radiology, meteorology, chemistry and economics) are carefully introduced in an approachable way to the readers who are interested in learning about the current literature on calibration methods.

There are many topics discussed in this volume, which consists of the nine chapters. More in detail, the primary research on the determination of the relationship between the input and output of an instrument or the essential value of the measuring object is contained in Chapter 1, 4, and 9. The recent research on the adjustment and/or compensation for the difference between the input/output model (calibration model) of the instrument and the measuring value is discussed in Chapter 2-3, and 5-9. The autonomous calibration method using an appropriate specimen is presented in Chapter 2, and 6, while the highly accurate determination method of the parameters by global optimization methods are introduced in Chapter 3, 5-6, and 9.

Chapter 8 offers a survey of the calibration methodology in the field of chemistry, while the remaining chapters present novel research results related to the calibration method applied in different fields. The summary of each chapter is as follows.

Chapter 1: Calibration of CMM (Coordinate Measuring Machine)

The strict definition of the calibration in metrology field and the calibration method of geometric error of CMM are introduced. In particular, a direct forward method to derive reliability indication on geometric error calibration of CMM by expanding propagation of errors and a geometric error model packed in a single linear equation with projection model for practical error evaluation possibilities are proposed.

Chapter 2: Calibration of three displacement sensors used surface straightness measuring system

An autonomous calibration method in which the zero-difference between three displacement sensors is calibrated by using the appropriate disc gauge (non-standard) is introduced.

Chapter 3: Calibration of a large-scale two-dimensional digitizer system

An on-site calibration method of a two-dimensional large-scale laser digitizer is proposed with the linearized coordinate calculation algorithm whose parameters are determined by the conventional global optimization method to get the better coordinate accuracy.

Chapter 4: Calibration of a nuclear radiation detection system

Calibration of a radiation measurement system relatively to two measurands, which are the ray energy and the radionuclide activity in the multi-group approximation, is proposed with uncertainty assessment based on the strict definition of calibration in the international vocabulary of metrology (VIM) as is mentioned before.

Chapter 5: Calibration of a rainfall-runoff model using a hybrid SCE-UA-KNN optimization algorithm

The calibration of a rainfall-runoff model is proposed with the hybrid SCE-UA-KNN global optimization method that is the Shuffled Complex Evolution of the University of Arizona (SCE-UA) optimization algorithm with the k-Nearest Neighbors (KNN) considering as a learning technique based on classifying objects according to the closeness to their K neighbors in a feature space, where K is the fixed number of neighbors.

Chapter 6: Calibration of a hydrological model using data depth function

The concept of calibration and validation using hydrological model is explained in detail. The calibration of a hydrological model is to find its optimal model parameters, but it is very complex global optimization problem as hydrological processes are non-linear processes. Therefore, the ROPE (Robust Parameter Estimation) algorithm which is one of global optimization algorithms using data depth function is mainly proposed to accurately determine the optimal parameter of the model.

Chapter 7: Calibration transfer of PLS (partial least squares) models using virtual standards in the chemistry

Applications of the use of virtual standards in transferring PLS calibration models which relate fuel properties to NIR (near-infrared) spectra are proposed. The virtual standards are digitally constructed using only a few neat chemicals acquired on the master and slave instruments.

Chapter 8: Review of the calibration methods in the chemistry

The key concepts of zero, first and second order calibration methods, in which multivariate calibration is divided regarding data complexity or dimensionality, are reviewed. And the application of these concepts by means of practical examples in pharmaceutical and inorganic analysis and polymeric materials is demonstrated.

Chapter 9: Calibration of a macroeconomic model with a Monte Carlo methodology

A Monte Carlo calibration methodology, assuming parameter uncertainty and prior distributions for the parameters, is applied in a macroeconomic model with labor income risk, incomplete markets, and heterogeneous agents. The determinants of precautionary savings and wealth inequality are studied in depth.

Since the topics in this volume are intended for a very broad audience, it is hard to provide a unified outlook. The book spans many professional fields where the calibration method is applied in different ways. Due to space limitation, for additional background readings on a specific field the reader is referred to the references included at the end of each Chapter.

Ikumatsu Fujimoto and Kunitoshi Nishimura

ACKNOWLEDGMENT

We are very grateful to Dr. Toshiyuki Takatsuji, National Metrology Institute of Japan (NMIJ), National Institute of Advanced Industrial Science and Technology AIST, who provided valuable advice during the course of this work. And also we are thankful to all authors in this volume for their cooperation and especially to Professor Marco Cozzi of Queen's University, one of authors, who corrected the preface and book description of this volume friendly and carefully. Finally, I (Ikumatsu Fujimoto) wish to express my deepest gratitude to my great teacher Dr. Mun Son-Myong.

In: Calibration Technology, Theories and Applications ISBN: 978-1-62618-808-2
Editors: Ikumatsu Fujimoto and Kunitoshi Nishimura

Chapter 1

RELIABILITY IN CALIBRATION OF GEOMETRIC ERROR OF COORDINATE MEASURING MACHINE

Makoto Abe*

National Metrology Institute of Japan,
National Institute of Advanced Industrial Science and Technology

ABSTRACT

Calibration in coordinate metrology is recognized as a challenging theme for development due to its nature having inherent complexity, instead significant usefulness in industry. Calibration of CMM may have an aspect to be emphasized that is establishment of estimation methodology of measurement uncertainty associated with respective measurands determined on workpieces to be measured. Along with it, this paper describes, firstly modeling of geometric errors based on quasi rigid body kinematics which elaborates one of fundamental characteristics of CMMs today, with measurement model which is independently applicable depending on practical situation. Secondly, an uncertainty estimation methodology on calibration of CMM is discussed. The methodology based on expansion of propagation of errors through the proposed CMM error model is presented.

Keywords: calibration, coordinate metrology, uncertainty, reliability

* Corresponding Author address: 1-1-1, Umezono, Tsukuba, Ibaraki, 305-8568, Japan
Email:Makoto.abe@aist.go.jp

1.1 INTRODUCTION

1.1.1 Coverage of Calibration

A term calibration seems to be used in various kinds of fields, however sometimes with varied interpretation. For example, calibration is often loosely interpreted or used as compensation or adjustment of instruments. Stringently examining it as a term used in metrology field, definition of what calibration means is described in the international document VIM (International Vocabulary of Metrology) [1] published by Joint Committee for Guide in Metrology (JCGM) organized by eight international organizations, BIPM, IEC, IFCC, ISO, IUPAC, IUPAP, OIML, and ILAC. According to VIM, calibration is an operation to establish a relation between intended values with measurement uncertainties provided by measurement standards and corresponding indications with associated measurement uncertainties. And this information can be used to establish a relation for obtaining a measurement result from an indication. It should be noted that the former part of the process alone can be performed as a calibration.

With intention to clarify scope and coverage of this article, definitions of important terms are listed below.

(1) Calibration

Calibration is defined as an operation to establish a relation between intended values to be obtained with measurement uncertainties provided by measurement standards and corresponding indications with associated measurement uncertainties. And this information can further be used to establish a relation for obtaining a measurement result from an indication.

(2) Measurement uncertainty

Measurement uncertainty is a parameter characterizing the dispersion of the quantity values attributed to a measurand.

(3) Metrological traceability

Metrological traceability is a property of a measurement result where the result can be related to the measurement unit through a documented unbroken chain of calibration.

Unit of length metre has been defined by length which light in vacuum flies through during one by 299792458 of a second since 1983 as one of a fundamental unit within International System of Units. The unit of meter is therefore fixed to unit of time and speed of light. However, practical realization of the metre may be matter of uncertainty required in the use mostly because of difficulty in realization of the definition of metre. Wavelengths of various stabilized lasers that have been recommended for use by international consensus. The iodine stabilized Helium-Neon laser with the fractional uncertainty of in the order of 10^{-11} is one of them that can be used to realize the unit of length as shown in Figure 1.

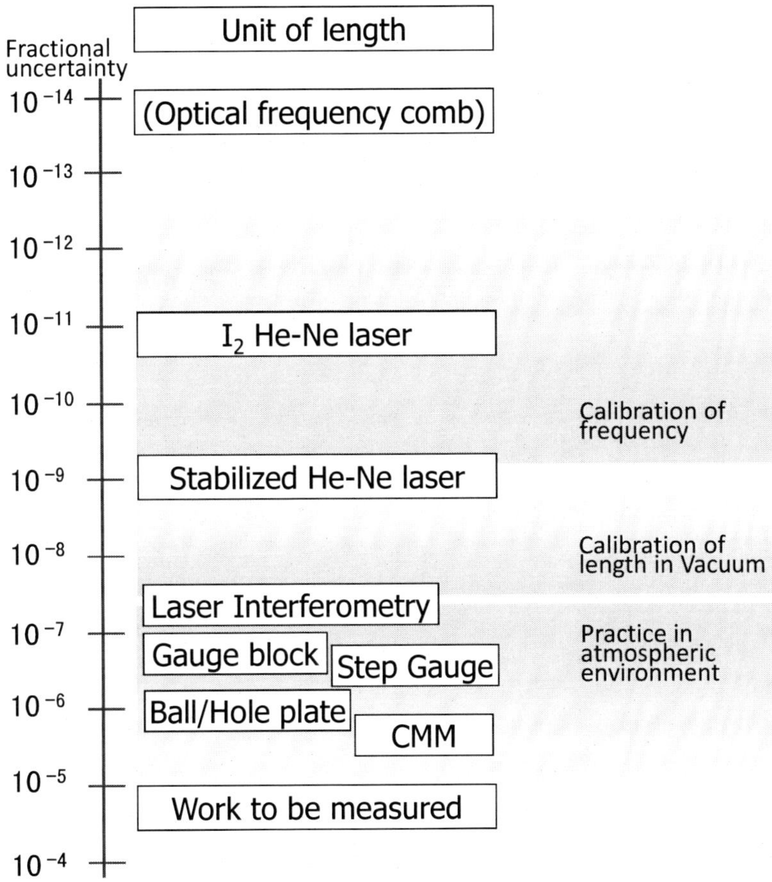

Figure 1. Traceability chain to unit of length

Most commonly used measurement standard to realize unit of length in practical use may be, gauge blocks, step gauges, and laser interferometers. These measurement standards may realize the fractional uncertainty of in the order of 10^{-7} in a typical laboratory environment.

Considering a practical situation for calibrating a geometrical measuring instrument such as CMM, one will have to refer a measurement standard which may be represented by an artifact somewhat calibrated in prior for having had the calibration values and the associated uncertainties. However a quantity to be calibrated on the CMM is preferred to be identical to that of the calibrated artifact, it is usually difficult to realize the situation. This is because CMM has indication in three dimensional space although typical calibrated artifact available in reality has one dimensional or sometimes two dimensional characteristics.

This is a reason why acceptance test and reverification test for CMM is performed by referring one dimensional calibrated artifact, such as gauge blocks or step gauges. Benefit of this approach is that the test result may have direct traceability to unit of length, and the test is relatively easy to perform for example for users of CMMs. On the other hand, it is not direct forward for testers to infer measuring performance of the CMM in three dimensional space. A motivation of this study comes from this point, which is trying to establish a calibration methodology in three dimensional space by referring one dimensional or two dimensional calibrated artifacts.

1.1.2 Complex case of calibration in coordinate metrology

Although CMM does not satisfy Abbe's principle and it may have strict limitation in improvement of accuracy or throughput, coordinate measuring machine (CMM) has been widely used in industry. The fact can be explained by following characteristics of CMM.

1) arbitrary measurement strategy can be designed by composing discrete measurement points,
2) arbitrary measurement datum required by design drawing of the part can be configured, although the datum can be independent to that of measuring machine.
3) CMM equipped with computer numerical control (CNC) is suitable for automation of measuring process.
4) Throughput of the measuring process has been drastically improved.
5) Rather higher accuracy about 1 micrometer for 1 m of size to be measured has been reached.
6) Various kinds of probing systems, not only a contacting probing system but also non contacting probing systems that based on optical sensing technologies, have been realized.

Especially characteristics described in 1) and 2) may potentially benefit wide range of CMM users in industry to utilize CMM for inspecting parts with complex geometrical tolerance and the related datum. Since digitization of technologies within production process rapidly diffuses importance of CMM in inspection of parts will further increase.

Measurement today, including geometrical measurement, is expected to demonstrate its results to conform to the corresponding metrological unit. In case of dimensional metrology the unit is the definition of metre. Calibration of instruments is recognized as a prerequisite to be able to do so.

1.2. Calibration of CMM based on Parametric Error

1.2.1 Geometric error modeling of CMM

Modeling of geometric errors of CMMs and corresponding compensation techniques have been studied for decades. Typically two different approaches have been realized namely:

1) Geometric errors of CMM at the end effector to be decomposed into translational and rotational errors each having three degrees of freedoms are mapped in grid wise in measuring volume.

2) Geometric errors of CMM at the end effector to be decomposed into a combination of guide ways, respective guide ways to have translational and rotational errors each having three degrees of freedoms.

Although the approach 1) may require certain amount storage volume such as a few of ten megabytes, by supposing an example of a CMM having 1 x 1 x 1 m^3 in the measurement volume, and having values for error map with 10 mm spacing, resulting 6 x 10^{2x3} =6 x 10^6 mapping points, the amount of error data itself does not give seriously negative impact due to significant processing capability of computer systems in recent years.

The approach 2) on the other hands may describe geometric errors of CMM with much lesser numbers of independent parameters, by supposing an example of a CMM having 1 x 1 x 1 m^3 in the measurement volume, and having values for error map with 10 mm spacing, resulting 18 x 10^2 =1800 mapping points. The characteristic of possibility to model geometric error of CMM by a linear combination of unknown variables is in line with interest of this study.

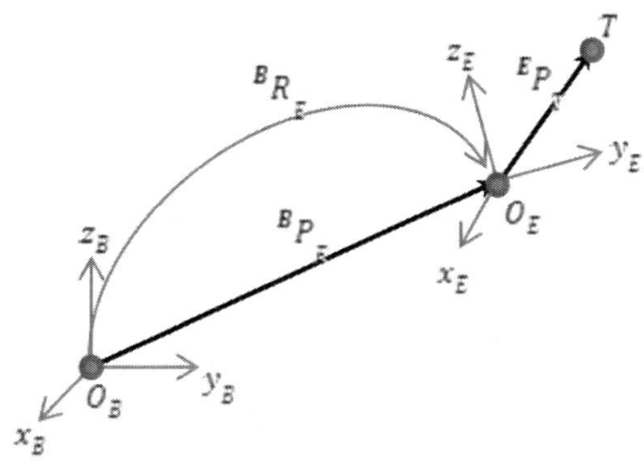

Figure 2. Coordinate transformation from base coordinate system to end effector

Let's consider position and orientation $^B X_T \in R^{6\times1}$ of an end effector T on the coordinate system 0_E of a CMM by relying on forward kinematics which starts its description from the base coordinate system 0_B. By having an orientation transformation matrix $^B R_E \in R^{B\times B}$, it yields equation (1). The schematic is shown in **Figure 2**.

$$^B X_T = \begin{bmatrix} ^B P_E \\ 0_{B\times1} \end{bmatrix} + \begin{bmatrix} ^B R_E & 0_{B\times B} \\ 0_{B\times B} & ^B R_E \end{bmatrix} \begin{bmatrix} ^E P_T \\ 1_{B\times1} \end{bmatrix} \tag{1}$$

Where, a superscript on the left shoulder of a symbol specifies coordinate system which the symbol quantity is based on. Equation 1 can be expressed as equation (2) by introducing homogeneous transformation matrix $^B T_E \in R^{4\times4}$,

$$^{B}X_{T}=\begin{bmatrix} ^{B}T_{E} & 0_{4\times4} \\ 0_{4\times4} & ^{B}S_{E} \end{bmatrix}\begin{bmatrix} ^{E}P_{T} \\ 1 \\ ^{E}P_{T} \\ 1 \end{bmatrix} \tag{2}$$

Where,

$$^{B}T_{E}=\begin{bmatrix} ^{B}R_{E} & ^{B}P_{E} \\ 0_{1\times3} & 1 \end{bmatrix},\ ^{B}S_{E}=\begin{bmatrix} ^{B}R_{E} & 0_{3\times1} \\ 0_{1\times3} & 1 \end{bmatrix} \tag{3}$$

When more than two joint components exist, equation (3) can be extended as it includes multiple coordinate transformations as equation (4).

$$^{B}X_{T}=\begin{bmatrix} ^{B}T_{1} & 0 \\ 0 & ^{B}S_{1} \end{bmatrix}\begin{bmatrix} ^{1}T_{2} & 0 \\ 0 & ^{1}S_{2} \end{bmatrix}\begin{bmatrix} ^{2}T_{1} & 0 \\ 0 & ^{2}S_{1} \end{bmatrix}\begin{bmatrix} ^{E}P_{T} \\ 1 \\ ^{E}P_{T} \\ 1 \end{bmatrix} \tag{4}$$

Geometric errors of CMM is typically small enough for us to be able to express the error model by linearly combining them for nominal geometry $X_{nominal}$ and that for deviational error δX, namely,

$$^{B}X_{T}=\begin{bmatrix} X_{nominal}+\delta X \\ 1 \end{bmatrix} \tag{5}$$

In case of calibrating a Cartesian CMM, the nominal model is usually composed of unity transformation. Calibration of geometric error of CMM can be said to be identification of δX as one of the possibilities.

The deviational error model of CMM with three translations and three rotations at the end effector of the CMM can be described as equation (6).

$$\delta X = KF \tag{6}$$

Where, respective parametric errors are linearly combined by the transformation matrix ***M***.

$$F=\begin{bmatrix} f_x & f_y & f_z & f_p \end{bmatrix}^T$$

$$f_x=\begin{bmatrix} txx & tyx & tzx & rxx & ryx & rzx \end{bmatrix}$$

$$f_y=\begin{bmatrix} txy & tyy & tzy & rxy & ryy & rzy \end{bmatrix}$$

$$f_z=\begin{bmatrix} txz & tyz & tzz & rxz & ryz & rzz \end{bmatrix}$$

$$f_p=\begin{bmatrix} txp & typ & tzp & rxp & ryp & rzp \end{bmatrix}$$

$$K = \left[\begin{pmatrix} I & {}^{x}R_{y} \\ 0 & {}^{x}R_{y} \end{pmatrix} \begin{pmatrix} I & {}^{y}R_{z} \\ 0 & {}^{y}R_{z} \end{pmatrix} \begin{pmatrix} I & {}^{z}R_{p} \\ 0 & {}^{z}R_{p} \end{pmatrix} \begin{pmatrix} I & {}^{p}R_{E} \\ 0 & {}^{p}R_{E} \end{pmatrix} \right]$$

Within these expressions, *tyx* denotes translation error in y-direction described as function of x-position, usually called as straightness, and *ryz* denotes rotation error in y-direction as function of z-position, usually called as yaw. ***F*** is a vector composed of a series of unknown parameters expressing respective parametric errors, such as translation errors and rotation errors of CMM.

Each parametric error can be expressed by various kinds of polynomial functions, provided the polynomial function is composed of a linear combination of unknown parameters and known basis functions. Following is an example of the expression a parametric error *tyx* by using B-spline function.

$$tyx(x) = \sum_{0}^{n-1} b_{i,k}(x) a_i \tag{7}$$

Where,

$$b_{i,k}(x) = \frac{x - q_i}{q_{i+k-1} - q_i} b_{i,k-1}(x) + \frac{q_{i+k} - x}{q_{i+k-1} - q_{i+1}} b_{i+1,k-1}(x) \tag{8}$$

$$b_{i,1}(x) = \begin{cases} 1 & (q_i \le x \le q_{i+1}) \\ 0 & (x < q_i, x \le q_{i+1}) \end{cases} \tag{9}$$

A methodology for linearly formulating geometric error model for CMM calibration is discussed. Solving equation (6) by referring known calibrated values delivered with a calibration artifact, all unknown parameters expressing CMM geometry in the form of parametric errors can be identified.

1.2.2 Modeling of measurement in practice

A typically available calibration artifact used for calibration of CMM may not provide information with 6 degree of freedoms simultaneously. One needs another formulation to correlate the above CMM model and available measurement technology in practice. On the other hand since the geometric error model does not contain quantity which is measurement technology specific. This characteristic makes it possible to adopt almost any kind of measurement technology.

1.2.2.1 One dimensional length measurement

One dimensional length measurement has been widely used for calibration of gauges and instruments. Most straight forward benefit of the one dimensional length measurement is to be able to obtain measurement result having direct traceability to unit of length the metre.

Recognizing this benefit, current ISO standard for testing CMM performance [2] utilizes one dimensional length measurement.

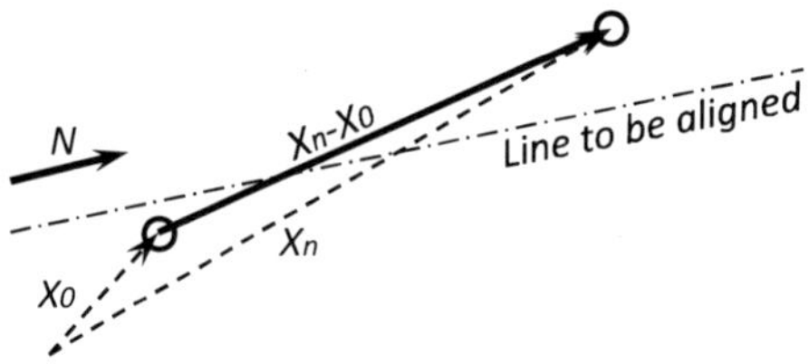

Figure 3. Projection model of one dimensional length measurement

Since one dimensional length measurement has sensitivity in the direction where the artifact is aligned to, appropriate projection operation is applied to project six degrees of freedom modeled in the previous section. **Figure 3** schematically shows relation between quantities to perform the projection. Projected longitudinal length deviation δl_{l0} observed by comparing calibrated length of the one dimensional artifact and indication of CMM is expressed.

$$\begin{aligned}\delta l_{l0} &= l_{cal} - l_{predicted} \\ &= l_{cal} - \left(\left(X_{nom} + X_{dev} - X_0 \right)^T N + c_0 \right) \\ &= l_{cal} - \left(\left(X_{nom} + e^T + \left(r \times p \right)^T - X_0 \right)^T N + c_0 \right)\end{aligned} \tag{10}$$

Since translation error of a CMM at the end effector is denoted by $e \in R^{3\times1}$, and the rotational error is done by $r \in R^{3\times1}$, both to be sub component of ${}^{B}X_T$ as below.

$${}^{B}X_T = \begin{bmatrix} e^T & r^T & 1 \end{bmatrix}^T$$

It yields,

$$\begin{aligned}\delta l_{l0} &= l_{cal} - \left(X_{nom} - X_0 \right)^T N - \left(N^T e + \left(p \times N \right)^T r + c_0 \right) \\ &= l_{cal} - \left(X_{nom} - X_0 \right)^T N - \begin{bmatrix} N^T & \left(p \times N \right)^T & c_0 \end{bmatrix} \begin{bmatrix} e \\ r \\ 1 \end{bmatrix}\end{aligned} \tag{11}$$

Where,

l_{cal} : one dimensional length measurement performed on a calibrated length artifact,

$l_{predicted}$: one dimensional length measurement obtained by using a CMM,

X_{nom} : nominal indication of the CMM,

X_{dev} : geometric deviation of the CMM,

X_0 : initial point of one dimensional measurement,

N : direction of one dimensional measurement,

c_0 : unknown offset due to positioning of calibrated artifact

p : probe offset vector

Equation (11) can be applied to one dimensional length measurement on CMM in any direction and orientation in the volume for any kind of one dimensional artifact such as gauge block, step gauge, laser interferometer, and so on.

1.2.2.2 Straightness measurement

Straightness measurement is also common in the field of dimensional metrology. The measurement is typically performed by referring a mechanically finished artifact having straight edge or flat face, or by laser interferometer. Commonly adopted numerical constraint for straightness excludes the averaged inclination of the measurement result. Here in two unknown offsets c_0 and c_1 due to positioning of an artifact for straightness measurement are combined in the model of straightness measurement. Projected transversal length deviation δl_{tr} observed by comparing calibrated straightness of the one dimensional artifact and indication of CMM is expressed.

$$\begin{aligned}\delta l_{tr} &= l_{cal} - l_{predicted}\\ &= l_{cal} - \left(\left(X_{nom} + X_{dev} - X_0\right)^T M + c_0 + c_1\left(X_{nom} - X_0\right)^T N\right)\\ &= l_{cal} - \left(\left(X_{nom} + e^T + (r \times p)^T - X_0\right)^T M + c_0 + c_1\left(X_{nom} - X_0\right)^T N\right)\end{aligned} \tag{12}$$

It yields equation (13) accordingly,

$$\begin{aligned}\delta l_{tr} &= l_{cal} - \left(X_{nom} - X_0\right)^T M - \left(M^T e + (p \times M)^T r + c_0 + \left(X_{nom} - X_0\right)^T N\right)\\ &= l_{cal} - \left(X_{nom} - X_0\right)^T M - \begin{bmatrix} M^T & (p \times M)^T & 1 & \left(X_{nom} - X_0\right)^T N \end{bmatrix} \begin{bmatrix} e \\ r \\ c_0 \\ c_1 \end{bmatrix}\end{aligned} \tag{13}$$

Where,

l_{cal} : one dimensional length measurement performed on a calibrated length artifact,

$l_{predicted}$: one dimensional length measurement obtained by using a CMM,

X_{nom} : nominal indication of the CMM,

X_{dev} : geometric deviation of the CMM,

X_0 : initial point of one dimensional measurement,

N : direction of one dimensional measurement,

M : direction of straightness error,

c_0 : unknown offset due to positioning of calibrated artifact,

c_1 : unknown offset due to orientation positioning of calibrated artifact,

p : probe offset vector

Equation (13) can be applied to straightness measurement on CMM in any direction and orientation in the volume.

1.2.2.3 Measurement using three dimensional artifact

Having three dimensional artifact which realizes volumetric collection of error of indication of CMM in three dimensional manner, calibration of CMM can be done. Deviational error vector δX can be calculated by comparing coordinate vector of the calibrated artifact X_{cal} and coordinate vector of the CMM X'_{CMM} on common coordinate space.

$$\delta X = X_{cal} - X'_{CMM} \tag{14}$$

Knowing information of nominal coordinate transformation to align the three dimensional artifact in coordinate system of the CMM, coordinate vector of the CMM on the coordinate system of the artifact X'_{CMM} becomes as below by introducing the nominal rotation matrix R_{nom} and the nominal translation vector T_{nom}.

$$X'_{CMM} = R_{nom} X_{CMM} + T_{nom} \tag{15}$$

Further introducing unknown parameters representing residual coordinate transformation by R_{res} for rotation transformation and T_{res} for translation transformation, it yields,

$$X'_{CMM} = R_{res}\left(R_{nom} X_{CMM} + T_{nom}\right) + T_{res} \tag{16}$$

Where, R_{res} can be linearized and modified, provided the residual coordinate transformation is relatively small in magnitude.

$$\begin{aligned} R_{res} &= \begin{pmatrix} 1 & -\theta_z & \theta_y \\ \theta_z & 1 & -\theta_x \\ -\theta_y & \theta_x & 1 \end{pmatrix} \\ &= I + \begin{pmatrix} 0 & -\theta_z & \theta_y \\ \theta_z & 0 & -\theta_x \\ -\theta_y & \theta_x & 0 \end{pmatrix} \\ &= I + \delta R_{res} \end{aligned} \tag{17}$$

Substituting R_{res} in equation (16) for that of equation (17), following equation is obtained.

$$\begin{aligned} X'_{CMM} &= \left(I + \delta R_{res}\right) R_n \left(X_{nom} + X_{error}\right) + \left(I + \delta R_{res}\right) T_{nom} + T_{res} \\ &\approx \left(R_{nom} X_{nom} + T_{nom}\right) + R_n X_{error} + \delta R_{res} \left(R_{nom} X_{nom} + T_{nom}\right) + T_{nom} \\ &= X'_{nom} + R_{nom}\left(I - U_{(P)}\right)\begin{bmatrix} e \\ r \end{bmatrix} + U_{(X'_{nom})}\theta + T_{res} \end{aligned} \tag{18}$$

Where,

$$\begin{aligned} X_{err} &= e + r \times P \\ &= \left(I - U_{(P)}\right)\begin{bmatrix} e \\ r \end{bmatrix}, \end{aligned}$$

and,

$U_{(P)}$: 3×3 matrix satisfying $U_{(P)} r = P \times e$

$$\begin{aligned} \delta R_{res} X'_{nom} &= \begin{pmatrix} 0 & -\theta_z & \theta_y \\ \theta_z & 0 & -\theta_x \\ -\theta_y & \theta_x & 0 \end{pmatrix} \begin{pmatrix} X'_{nx} \\ X'_{ny} \\ X'_{nz} \end{pmatrix} \\ &= \begin{pmatrix} 0 & X'_{nz} & -X'_{ny} \\ -X'_{nz} & 0 & X'_{nx} \\ X'_{ny} & -X'_{nx} & 0 \end{pmatrix} \begin{pmatrix} X'_{nx} \\ X'_{ny} \\ X'_{nz} \end{pmatrix} \\ &= U_{(X'_{nom})}\theta \end{aligned}$$

It yields,

$$\delta X = X_{cal} - X'_{nom} - \begin{bmatrix} R_{nom}\left(I - U_{(P)}\right)K & U_{(X'_{nom})} & I \end{bmatrix} \begin{bmatrix} \begin{bmatrix} e \\ r \end{bmatrix} \\ \theta \\ T_{res} \end{bmatrix} \tag{19}$$

Equation (19) can be applied to three dimensional calibration artifact for CMM in any direction and orientation in the volume.

1.2.2.4 Measurement using two dimensional artifact

The measurement model for a two dimensional artifact is almost identical to that of a three dimensional artifact, due to the fact that degrees of freedoms needed to perform the alignment are three dimensional. The difference in typical case is that one of three measured response variable in the direction of principal normal vector of the two dimensional artifact is not used for calibration.

1.2.3 Uncertainty evaluation in calibration of CMM

CMM can be categorized as a geometrical measuring system performing various measuring tasks based on a series of discrete measurement points obtained in three dimensional space. The characteristic enables CMMs to adapt high flexibility in designing measurement strategy, i.e. allocation of discrete points for the measurement task, and position and orientation of the workpiece including even that of practical realization of datum. Therefore measurement uncertainty of geometrical features on a workpiece may drastically be varied depending on how measurement strategy is determined. Accordingly measurement uncertainty in calibration of CMM may be varied due to design of calibration strategy.

One of the motivations of adopting simultaneous linear equations to perform calibration of geometric error of CMM is to evaluate uncertainty of calibration by utilizing the propagation of error.

Given the observation equation as equation (6), variance of estimated parameters $\widehat{F}$ will be calculated by propagation of error by equation (20). Here S_Y is variance of observation value Y.

$$S_{\widehat{F}} = \left(K^T S^{-1}{}_Y K\right)^{-1} \tag{20}$$

Once variance of estimated parameters is calculated, variance of estimated value of observation $S_{\widehat{Y}(x=x_0)}$ can also be calculated as equation (21).

$$S_{\widehat{Y}(x=x_0)} = K_{(x=x_0)} S_{\widehat{F}} K^T_{(x=x_0)} \tag{21}$$

Knowing linear relation between components within variance-covariance matrix, the partial matrix expressing contribution to translation error and that to rotation error is described.

$$S_{tra\overline{(x=x_0)}} = \left(K_{Tra(x=x_0)}\right) \times \begin{pmatrix} S_{\hat{F}TxxTxx} & S_{\hat{F}TxxTyx} & \cdots & \\ S_{\hat{F}TyxTxx} & S_{\hat{F}TyxTyx} & & \\ \vdots & & \ddots & \\ & & & S_{\hat{F}TzzTzz} \end{pmatrix} \times \left(K_{Tra(x=x_0)}\right)^T \tag{22}$$

$$S_{rot\overline{(x=x_0)}} = \left(K_{rot(x=x_0)}\right) \times \begin{pmatrix} S_{\hat{F}RxxRxx} & S_{\hat{F}RxxRyx} & \cdots & \\ S_{\hat{F}RyxRxx} & S_{\hat{F}TyxRyx} & & \\ \vdots & & \ddots & \\ & & & S_{\hat{F}RzzRzz} \end{pmatrix} \times \left(K_{rot(x=x_0)}\right)^T \tag{23}$$

Accordingly, variance-covariance matrix of the estimated parametric error curve is described.

$$S_{Txx\overline{(x=x_0)}} = K_{Txx(x=x_0)} S_{\hat{F}TxxTxx} K_{Txx(x=x_0)}^{\ T} \tag{24}$$

$$S_{Rxx\overline{(x=x_0)}} = K_{Rxx(x=x_0)} S_{\hat{F}RxxRxx} K_{Rxx(x=x_0)}^{\ T} \tag{25}$$

Reliability of each parametric error curve or that of combination of them can be evaluated.

1.2.3.1 Modeling spatial constraint of error for generating trial values

Utilization of specification or the test result of ISO 10360-2 test as principal information for uncertainty evaluation is considered. The ISO standard is internationally accepted as an acceptance and reverification test procedure for CMM. The procedure consists of two major tests. One is E-test measuring one dimensional size with five different spacing over the whole measurement volume along seven different directions. Commercially available CMM conforming to this standard claims the specification in the form of the maximum permissible error (*MPE*). The test is so performed as to be able to show conformance to the specification by taking the test uncertainty in account. An example of a typical expression of the specification of the E-test can be e.g. $MPE_E = a + b \times l$. Applied the MPE_E on shorter size of l, it shows rather smaller deviation like a, while larger the size the expression allows deviation to enlarge correspondingly. This tendency falls basically in line with our daily experience, although deviation observed on a real CMM reveals random like fluctuation in combination with moderately varying wave form. The latter is mostly caused by auto reflection of the unknown systematic contribution. It is natural to grasp that the expression of the maximum permissible error varying positively with the size increment includes information of intrinsic spatial constraint of error of CMM.

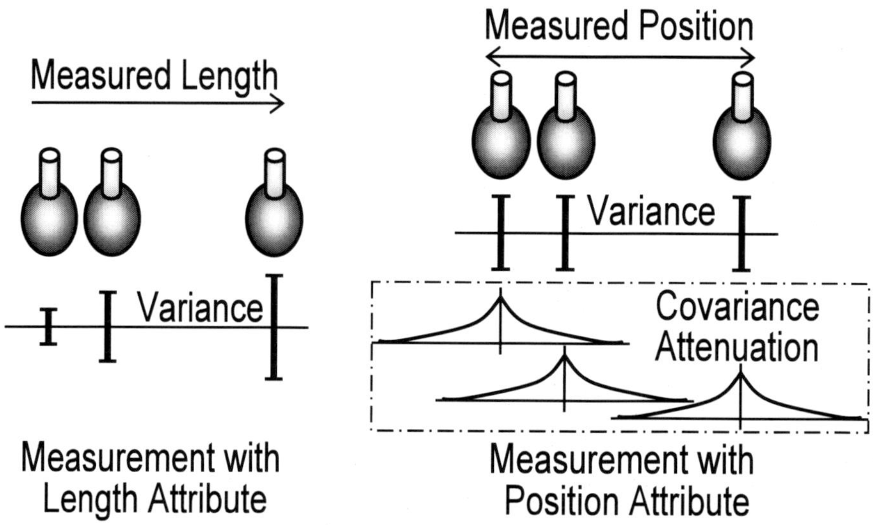

Figure 4. Modeling of spatial constraint in geometric error

A way of modeling is illustrated in Figure 4 by recognizing specification or the test result of ISO 10360-2 to be an indirect expression of uncertainty in size measurement. This model with maximum permissible sense is inspired by a fact that the specification and the interpretation of the test result shall be applied to any location and orientation in the CMM measurement volume.

1.2.3.2 Derivation of measurement strategy specific covariance matrix

A derivation procedure of task specific covariance matrix from specification or test result of ISO 10360-2 is described.

Considering a result of size measurement $l=x2-x1$ composed of two point coordinates $x1$ and $x2$, variance of the size measurement $Var(l)$ results equation (26) by introducing variance of respective point coordinates and the covariance $Cov(x1,x2)$ as the mutual influence. Variance of the point coordinate is assumed to be homogeneous in the measurement volume as the worst scenario. The mutual effect is explained as dependant of the distance.

$$\begin{aligned} \mathrm{Var}(l) &= \mathrm{Var}(x1)+\mathrm{Var}(x2)-2\mathrm{Cov}(x1, x2) \\ &= 2[\mathrm{Var}(x)-\mathrm{Cov}(l)] \end{aligned} \tag{26}$$

Let us simplify discussion here by interpreting MPE_E be comparable to 95% probability limit of normal distribution.

$$\mathrm{Var}(l)=\left[\left(a+b\times l_{\max}\right)/2\right]^2/2 \quad (27)$$

$$\mathrm{Cov}(l)=\left[\mathrm{Var}\left(l_{\max}\right)-\mathrm{Var}\left(l\right)\right]/2 \quad (28)$$

Where, l_{max} expressing physically accessible longest length in the measurement volume is introduced. Assuming homogeneous variance in the volume and the mutual influence attenuated depending on the distance, variance-covariance of respective point coordinates composing a complex measurement strategy can be quantified in straight forward manner.

1.2.3.3 Orthogonal decomposition of variance-covariance matrix

Having variance-covariance information, a unique Monte Carlo simulation fully reflecting the given statistical characteristics is performed. Suppose our having a measurement strategy composed of discrete point coordinates, e.g. n-points, a task specific variance-covariance matrix C with $n \times n$ dimension is assigned, and transferred to single run pre-process to decompose it orthogonally into the eigen vectors V and the corresponding eigen values D.

$$C = VDV^{-1} \quad (29)$$

Where,

$$V = \left[\left(V_1\right) \quad \left(V_2\right) \quad \cdots \quad \left(V_n\right)\right] \quad (30)$$

$$D = \begin{bmatrix} \lambda_1 & & & 0 \\ & \lambda_2 & & \\ & & \ddots & \\ 0 & & & \lambda_n \end{bmatrix} \quad (31)$$

The important point is that the eigen vector means orthogonal basis function and the eigen value does re-combination factor for our considering these information to be used for uncertainty evaluation on calibration of CMM. An example of the basis function in case of one dimensional size

Measurement with 100 mm equi-distanced measurement strategy is shown in Figure 5. A series of lower order exponential curve-like wave forms are observed.

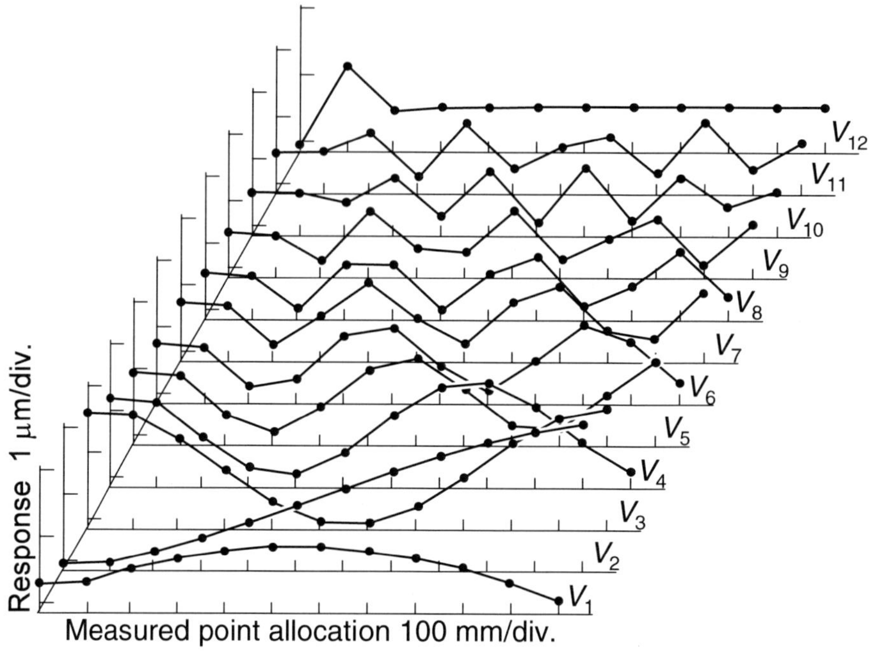

Figure 5. Example of orthogonally decomposed basis in case of one dimensional size measurement

1.2.3.4 Generating trial values reflecting variance-covariance

Once task specific basis functions and the corresponding re-combination factors are obtained, a string of trial values $\hat{x}$ can be generated easily.

$$\left(\hat{x}\right) = \varepsilon_1 V_1 + \varepsilon_2 V_2 + \cdots + \varepsilon_n V_n \tag{32}$$

Where, ε_i is a random number satisfying the expectation value to be zero and the variance to be equal to the corresponding eigen value. The simple equation (32) explains fact that repeated run evaluation for the task specific uncertainty evaluation can be performed by applying only a simple random generator.

1.2.4 Validation of calibration by geometric error model

The geometric error model proposed may provide quantitative statement for reliability of the calibration result. It expects statistical information from the observation data in the form of a covariance matrix formulated by variance and correlation information.

Although it is common for calibration bodies to state uncertainty in calibration measurement under commercial situation, the uncertainty is typically calculated under assumption of the possible worst scenario, such as with acceptable but with extreme environmental condition. Actual uncertainty can therefore always be better than that of claimed. This fact makes actually performed experimental validation of the geometric error model difficult. This study adopts validation by simulation to overcome the difficulty.

The simulation is configured by emulating real conditions observed on a real CMM. A conventional bridge driving CMM having measuring spindle at the end of the Z axis with the measurement volume of 500 x 400 x 400 mm^3 is considered. Totally about 130 parameters expresses 18 geometric errors. A hole plate measurement is chosen as a method to collect realistic geometric error in discrete distribution. Grid pitch of the hole plate to detect geometric errors of the CMM is chosen to be 50 mm which is conventional.

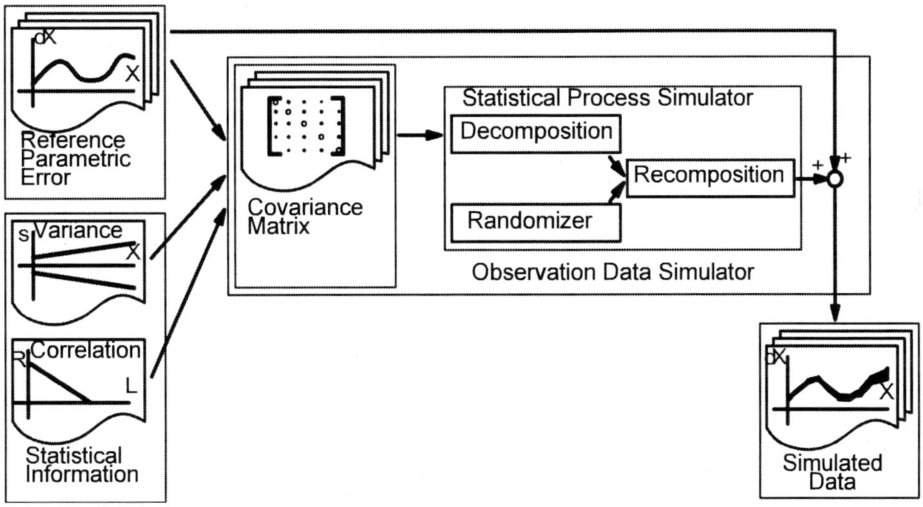

Figure 6. Schematic diagram of virtual hole plate measurement

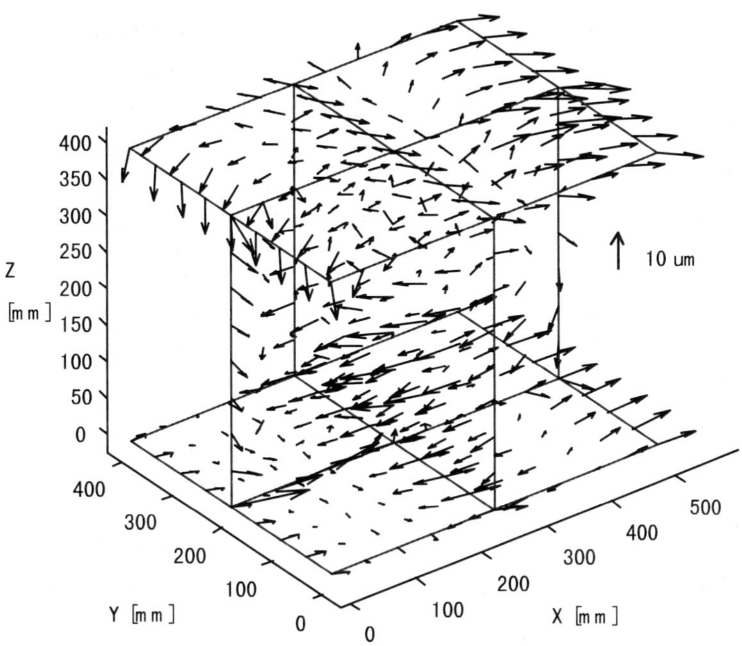

Figure 7. Error vectors of CMM detected by hole plate measurement

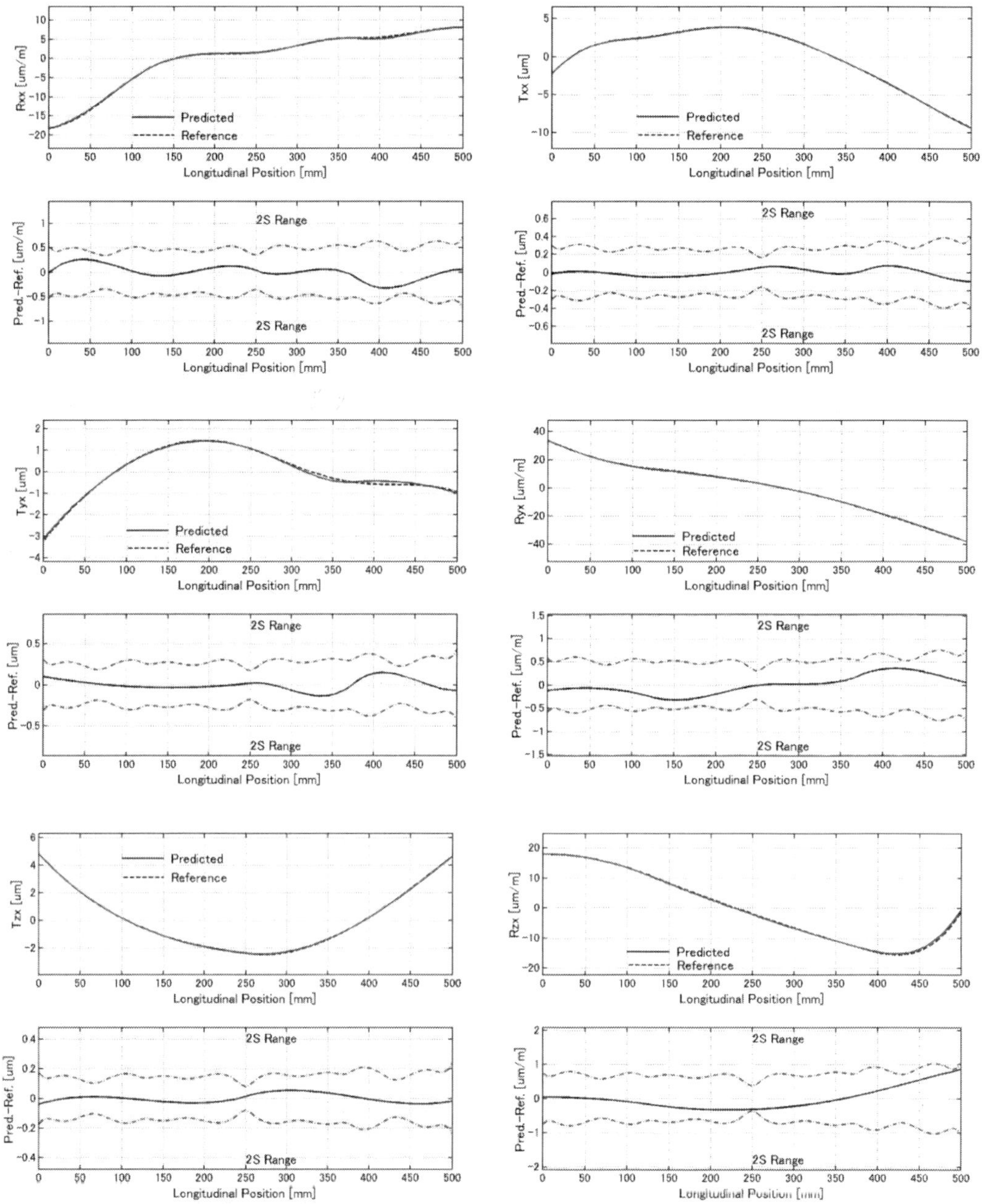

Figure 8. Example of estimation result of geometric errors and the associated reliability range with 95% probability

Reference values of geometric error calibration of the CMM are chosen to be that practically obtained on a real CMM with the same type. Observed geometric error by performing virtual hole plate measurement is composed by the procedure depicted in Figure 6. Reference geometric errors and calibration uncertainty of the hole plate are linearly combined and input to the simulation. Virtual hole plate measurement on the CMM is performed by four different principal hole plate position in the measurement volume of the

CMM. Figure 7 depicts position of hole plate and the observed error vector by virtually measuring the hole plate. A 200 mm probe extension is used to perform hole plate measurement on XZ plane and YZ plane.

To realize appropriate projection of observed errors to geometric error space of the CMM, the procedure described in the section Measurement using two dimensional artifact is applied.

Once a series of observed error is obtained by using the hole plate, estimation of geometric errors of the CMM is performed according to the model described by equation (6). Since four different hole plate positions are measured, especially with two different probe extensions in case of XZ plane and YZ plane configured, totally six hole plate measurement results are input for estimation of the geometric errors and the corresponding reliability statements. The expanded uncertainty of calibration of the hole plate is given by equation (33).

$$U\left[\mu\text{m}\right] = 0.6 + \frac{0.9}{1000} \times L\left[\text{mm}\right] \tag{33}$$

Examples of estimation results of the geometric errors are depicted in Figure 8 for the results related to the X-axis. Within the respective upper plots, solid line shows estimation result of the geometric error curve. The dotted line shows the reference curve given for generating observation values before the simulation. Respective lower plots show difference of two curves with reliability range of 95 %. Not only are the plots depicted here, but also those for all the geometric error curves, estimation results of geometric error curves show good conformance with the reliability ranges.

1.3. SUMMARY

A calibration methodology for coordinate measuring machine to be able to derive reliability indication in three dimensional space is described. The method is able to track complex error propagation process during calibration of CMM. A potential limitation of the method may be dimension of the variance-covariance matrix especially on application requiring a large number of discrete points. A possible solution is to section the matrix into plural parts where the mutual influence to be negligible.

REFERENCES

[1] International vocabulary of metrology –Basic and general concepts and associated terms (VIM), JCGM 200. BIPM, IEC, IFCC, ISO, IUPAC, IUPAP, OIML, and ILAC : 2008.

[2] Geometrical product specifications (GPS) -- Acceptance and reverification tests for coordinate measuring machines (CMM) -- Part 2: CMMs used for measuring linear dimensions. 2009.

[3] Uncertainty estimation using Monte-Carlo method constrained by correlations of the data, M. Nara, M. Abbe, K. Takamasu, Key Engineering Materials, Vols.381-382, pp.587-590, 2008.

In: Calibration Technology, Theories and Applications ISBN: 978-1-62618-808-2
Editors: Ikumatsu Fujimoto and Kunitoshi Nishimura

Chapter 2

AUTONOMOUS CALIBRATION METHOD OF THE STRAIGHTNESS MEASURING MACHINE WITH THREE-POINT METHOD AND ITS UNCERTAINTY ANALYSIS

***Ikumatsu Fujimoto*[*1] *and Kunitoshi Nishimura*[2]**
[1]Okuwa Technical Research Center, Yamaguchi, Japan
[2]Mizukoshi Keiki Co Ltd., Tokyo, Japan

ABSTRACT

An autonomous calibration method of the zero-difference (i.e. the difference between zero values of each sensor probe) in the three-point method using three displacement sensors for measuring the surface straightness is proposed with the uncertainty analysis. The three -point method has the advantage that the influence of the motion errors such as up-down movement error and pitching error occurred due to the scanning motion can be eliminated automatically. However, the zero-difference between the three sensors brings a serious error to the measurement values. Therefore, it has to be calibrated accurately.

In this research, a simple disc gauge made by CNC turning machine, etc. is used for the calibration of the zero-difference between the three sensors. The calibration method is comprised with two steps. In the first step, the disc gauge rotates a few revolutions and moves parallel to the three displacement sensors built into a holder. In the second step, geometrical parameters between the sensors and the disc gauge are simultaneously acquired at each predetermined movement of the sensors and the zero-differences are computed by a new algorithm.

This chapter is composed of the following three themes.

[1] The basic theory and algorithm for realization of the proposed method is explained with the uncertainty analysis and numerical simulation.

* Corresponding Author: E-mail : fujimoto.ikumatsu@gmail.com

[2] The relationship between the disc gauge radius and the measurement uncertainty is investigated.
Keywords: Autonomous calibration method, Zero-difference, Straightness measuring system, Displacement sensor, Three-point method, Motion error, Disc gauge, Uncertainty analysis

2.1. INTRODUCTION

In this chapter, we introduce one example of the calibration method with respect to the measuring system of the flatness of an object surface such as the silicon wafer surface in the semi-conductor manufacturing process. The meaning of the word "calibration" described in this chapter is to adjust an output of the measuring system to the essential value of the object, which is determined by using standard or standard specimen. The standard is an object that bears a defined relationship to a unit of measure (length, volume, plane and others) used for calibration of measuring devices. Further, a self-calibration or an autonomous calibration is to execute the above process automatically without the standard or the standard specimen, and one example is proposed in this chapter.

A straightness measuring system which is a scanning probe system to determine the straightness of the object surface is one of the important fundamental methods to measure the flatness of an object surface [1,2].

Such a system is generally realized as follow: (1) scan the object surface by a sensor holder with some displacement sensors and (2) measure the geometrical relationship between the sensors and the object surface at each predetermined movement of the sensors.

In all straightness measuring systems, the motion errors (i.e. up-down movement error and pitch error) during the scanning motion of sensor cause serious errors to the measured value and cannot be avoided. Because, as the scanning sensor holder moves along a guide rail which is not straight in a strict sense, the scanning motion inevitably invites the motion errors which are up-down movement error and pitch error, etc.

In one-point method using one displacement sensor, the motion errors are unavoidable. In case of two-point method using two displacement sensors [3-5], the up-down movement error can be eliminated by subtracting the measured value of the one sensor from that of the another sensor. However, the multi-point method using three or more displacements sensors built-in a holder [6-8], and the two-point method using two or more angle sensors [9] have been frequently utilized because these method can automatically eliminate the influence of both the motion errors mentioned above. However, these methods require more cost and that their accumulative error (i.e. a parabolic error with respect to the measured surface) caused by the iterative measuring which is required to determine the whole shape of an object surface could be larger than that of the one-point method or the two-point method which is adapted for the high accurate monitoring system which can neglect the motion errors. Furthermore, it is difficult for these systems to detect the high frequency components of the measuring object surface [10].

The multi-point method can reduce this accumulative error by making the sampling period of measuring the object surface shorter except the influence of systematic error elements [11].

Therefore, these methods are useful for users to measure an object surface because of high measuring accuracy and speed in case that the measuring area of an object surface is comparatively narrow and the measurement of high spacial frequency components is not necessary.

However, the zero-difference, which is the difference between the zero value of each sensor, brings a parabolic serious error to the measured value and its error becomes much larger as the length of object becomes bigger.

Therefore, it is the most important problem of the multi-point methods to determine the zero-difference accurately.

The typical method to determine the zero-difference is to adjust the zero-difference using a standard gauge to zero. But as the standard gauge is very expensive and a slight amount of the zero-difference invites large errors, the method cannot be used. Other method without using a standard gauge is the reversal method [12-14]. But this method is suitable only for measuring the straightness of cylinder. Furthermore, the methods [15,16] which use an autocollimator or an interferometer for the zero-difference calibration, by which high measuring accuracy can be realized, are seemed not to be practical because of high cost and low measuring speed. That is, an easy and conclusive practical method has not been developed yet.

In this chapter, an autonomous calibration method of the zero-difference in the three-point method for measuring the surface straightness is proposed [17-19]. A simple disc gauge made by CNC turning machine, etc. is used for calibration. The conceptual design and algorithm of the proposed method for realization are described with its uncertainty analysis in detail.

By the way, there is a method using an interferometer in addition to the scanning method above-mentioned to measure the flatness. However, the method is not referred in this chapter. Please see the reference [20], etc. if you want to know the autonomous calibration method in the interferometer system.

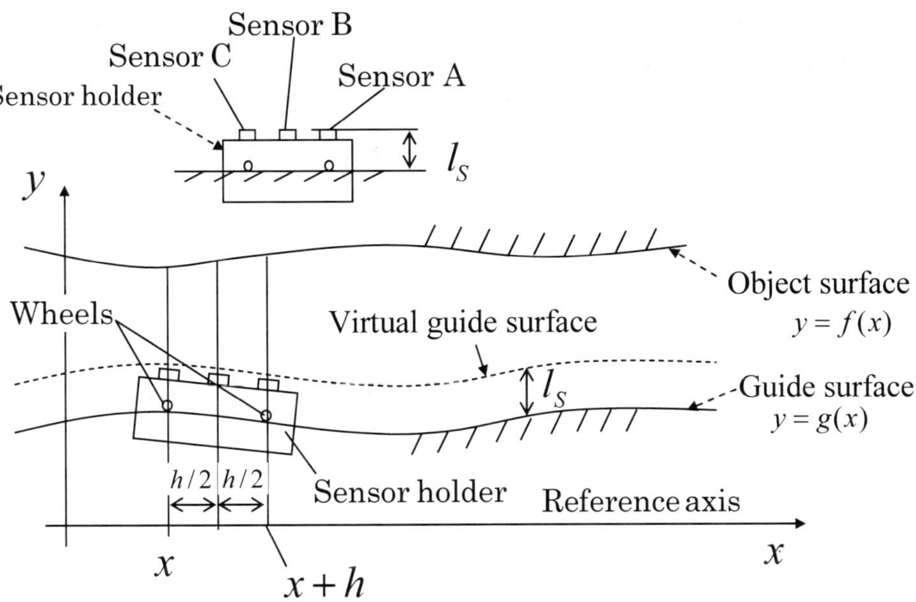

Figure 1. Schematic view of a three-point straightness measuring method.

2.2. Outline of the Straight Measuring System with Three-Point Methode and its Problems

In this section, an outline of the general algorithm of three-point method is introduced with the zero-difference problem.

Fig. 1 shows a three-point straightness measuring system whose sensor-holder contains three displacement sensors. The guide surface of the sensor holder and the object surface are roughly arranged in parallel to the reference axis x, respectively as shown by Fig. 1. The irregularity of the both surfaces are depicted to be very large, but supposed to be so small that it is invisible to the eye.

The three displacement sensors A, B, C are arranged at even intervals and their longitudinal axes are in parallel.

At a first, the three sensors are calibrated using a standard gauge before measuring so that each output data is expressed as

$$\mu_j(z) = C_j z + s_j \quad (j = A, B, C) \tag{1}$$

where variables s_j and C_j denote the unknown zero-values and sensitivity coefficients of sensors A, B, and C, respectively and also variable z defines the distance between the sensor top and an object surface.

At a next step, each zero-value is adjusted so that the following relation is satisfied,

$$S_B = (S_A + S_C)/2\,. \tag{2}$$

The relation of equation (2) can not be realized because the standard gauge surface is not flat in a precise mathematical sense. Therefore, zero-difference between three sensors is assumed to be expressed as

$$S_E = (S_A + S_C) - 2S_B\,. \tag{3}$$

An algorithm for measuring the straightness of an object surface is derived through the following process. At first, some variables are denoted by

[1] Shape of object surface: $f(x)$,
[2] Shape of guide surface: $g(x)$,
[3] Pitch error of sensor holder: $e_p(x)$,and
[4] Adjacent sensor interval: $h/2$.

Both the guide surface and the object surface are assumed to be roughly arranged in parallel to the reference line and the sensor holder is assumed to move on the guide face according to the guide line and that the positions of the two wheels are assumed to be mounted in the positions of the sensors A, C.

Therefore, each output of the sensors A, B and C can be expressed as

$$\mu_A(x) = C_A\left[f(x+h) - g(x)\right] + s_A, \tag{4}$$

$$\mu_B(x) = C_B\left[f(x+h/2) - g(x)\right] + s_B, \tag{5}$$

$$\mu_C(x) = C_C\left[f(x) - g(x)\right] + s_C. \tag{6}$$

The outputs are regarded as a function of x, as the distance between the sensor top and the object surface varies according to the scanning. Moreover, the equations are derived on the following supposition. First, the inclination of the sensor holder is neglected as it is very small. Second, the length l_S between the sensor top and the guide surface is neglected as it is constant. That is, the guide surface is supposed to be shifted to the virtual guide surface as shown in the Fig. 1, where the shifting value l_S is the distance between the sensor top and the guide surface.

So, the pitch error is shown by

$$e_P(x) = \left(g(x+h) - g(x)\right)/h. \tag{7}$$

Therefore, each output obtained from the sensors A, B, and C can be rewritten by

$$\mu_A(x) = C_A\left[f(x+h) - g(x) - e_P(x)\cdot h\right] + s_A, \tag{8}$$

$$\mu_B(x) = C_B\left[f(x+h/2) - g(x) - e_P(x)\cdot h/2\right] + s_B, \tag{9}$$

$$\mu_C(x) = C_C\left[f(x) - g(x)\right] + s_C. \tag{10}$$

Then, the second order differential of the object surface *f(x)* can be expressed by the output values $\mu_A(x)$, $\mu_B(x)$, and $\mu_C(x)$ as

$$f''(x) \cong M(x) - s_E/(h/2)^2, \tag{11}$$

where

$$f''(x) \cong \left\{f(x+h) - 2f(x+h/2) + f(x)\right\}/(h/2)^2, \tag{12}$$

$$M(x)=\frac{1}{(h/2)^2}\left\{\frac{\mu_A(x)}{C_A}-2\frac{\mu_B(x)}{C_B}+\frac{\mu_C(x)}{C_C}\right\}, \tag{13}$$

$$s_E=\frac{1}{(h/2)^2}\left(\frac{s_A}{C_A}-2\frac{s_B}{C_B}+\frac{s_C}{C_C}\right). \tag{14}$$

By integrating Eq. (11) two times, the object surface $f(x)$ can be approximately determined by the following equation.

$$f(x)=\int_0^x\int_0^s M(\alpha)d\alpha ds-2s_E(x/h)^2-C_1x-C_0\ \ (x>0) \tag{15}$$

where C_1 and C_0 are constant numbers.

As is shown by Eq. (15), the zero-difference brings a parabolic error $2s_E(x/h)^2$ to the measured object surface. For example, a value $2(x/h)^2$ is 800 when $x/h=20$. Therefore, the influence of the parabolic error on it becomes 800 times.

The accumulative uncertainty $\delta_{acc(n)}$ except systematic error elements of sensor output, which is caused by the sequential addition of the measuring values so as to determine a whole surface shape of the object, brings the uncertainty proportional to a $(3/2)^{th}$ order value of the measuring length l as is shown by Eq. (16).

$$\delta_{acc(n)}=\frac{4\delta_m}{h^2}\sqrt{2sl^3+3s^2l^2+s^3l}=\frac{4\delta_m}{h^2}\left(\sqrt{2\frac{1}{n}+3\frac{1}{n^2}+\frac{1}{n^3}}\right)l^2, \tag{16}$$

where

(1) δ_m : measurement uncertainty of each sensor, (2) $h/2$: adjacent sensor interval, (3) n : sampling number of measurement, and (4) s: sampling period of measurement.

However, the accumulative uncertainty can be reduced sufficiently small by making the sampling period of measurement shorter as shown by Eq. (16). In order to obtain better accuracy, the error element caused by the zero-difference from the measuring value should be subtracted. So a method to calibrate the zero-difference accurately is proposed.

2.3. Autonomous Calibration Method of Zero-Difference

2.3.1. Calibration of the Zero-Difference in Three-Point Method

A disc gauge made by CNC turning machine is used for the calibration of the zero-difference. As is shown in Fig. 2, the disc gauge rotates so that the middle sensor B scans relatively on a circumference of a circle with radius r and the other sensors A and C scan on about the same circumference which is different from that of a circle with radius r.

The sensor holder with the three sensors is assembled over the disc gauge so that its surface shape can be measured by the three sensors. The three sensors A, B (middle) and C

are positioned in a line which is perpendicular to the line that connects the rotational center point O and the middle sensor B and that is parallel with rotational axis as shown in Fig. 3 and 4.

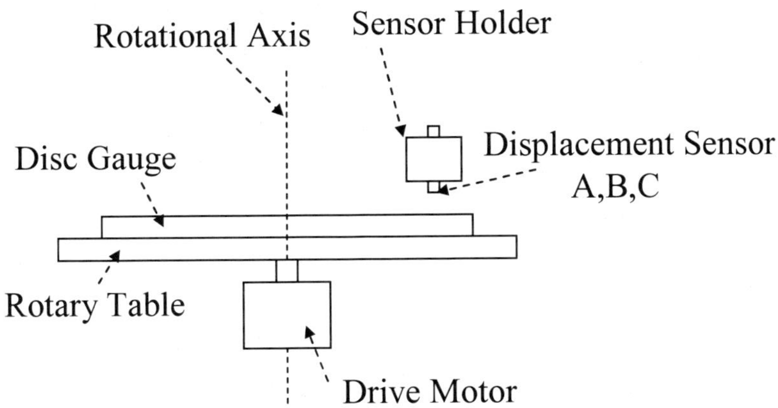

Figure 2. A conceptual design of the proposed calibration method.

We suppose that the clockwise rotating angle of disc gauge is θ, the radius of a circle which the middle sensor B scans is r, then the displacement x on the circumference of a circle with radius r is described by

$$x = r\theta . \tag{17}$$

The disc rotates around the rotating axis in a clockwise direction and the observation equations can be obtained from the outputs of the three sensors A, B and C, which are the same equations as Eqs. (8) - (10). Here, *f(x)* shows the surface of measuring object along the circle.

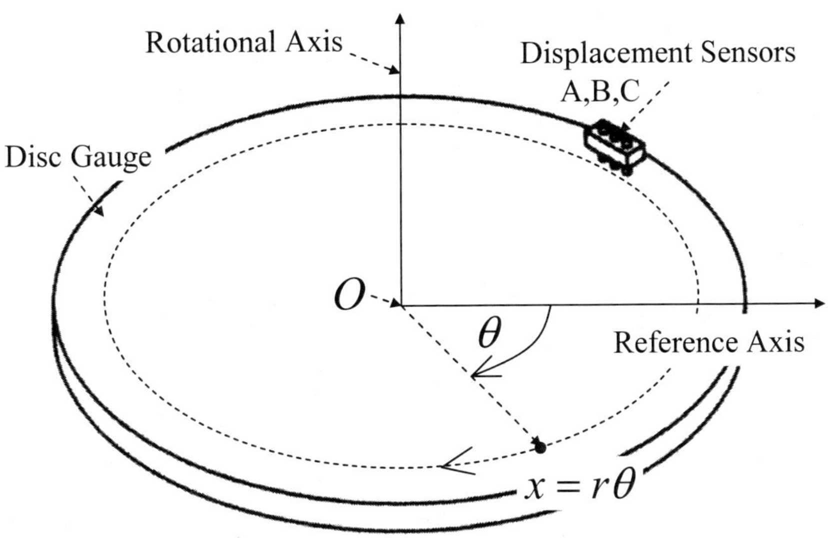

Figure 3. Perspective illustration of example configuration of proposed calibration method.

As the three sensors can never rotate on the same radius as shown in Fig. 4, it is an essential problem how accurately the behaviors of the three sensors A, B and C can be approximated by Eqs. (8) - (10). The error caused by the difference of rotating radius between sensor B and sensor A, C becomes one of the systematic errors in this measuring system. At first, we assume that the three sensors which are arranged in alignment can scan the circumference in same radius r.

When the disc rotates one revolution, the variable x on the circumference with radius r changes $L=2\pi r$. So, the line integral and double line integral of the numerical sequence $M_r(x)$ defined by Eq. (13) can be described by

$$\int_0^x M_r(s)ds = f'(x) + 4s_E(x/h^2) + C_1\,(x>0)\,, \tag{18}$$

$$\int_0^x \int_0^s M_r(\alpha)d\alpha ds = f(x) + 2s_E(x/h)^2 + C_1 x + C_0 \;\;(x>0)\,, \tag{19}$$

respectively.

Where

$$M_r(x) = \frac{1}{(h_r/2)^2}\left\{\frac{\mu_A(x)}{C_A} - 2\frac{\mu_B(x)}{C_B} + \frac{\mu_C(x)}{C_C}\right\} (h_r \text{ is denoted by Eq.(27)}) \tag{20}$$

As each position of the disc surface turns back to the same position after one revolution, the function $f(x)$ and its differential function $f'(x)$ have the following property, respectively.

$$f(0) = f(L)\,, \tag{21}$$

$$f'(0) = f'(L)\,. \tag{22}$$

Therefore, the zero-difference can be easily obtained by Eqs. (17) and (18) as

$$s_E = \frac{h^2}{4L}\oint_{0\le x\le L} M_r(x)dx = \frac{h^2}{8\pi}\oint_{0\le\theta\le 2\pi} M_r(r\theta)d\theta\,, \tag{23}$$

because the following relation can be approved by Eq. (18) and (22).

$$\oint_{0\le x\le L} M_r(x)dx = f'(L) + 4s_E(L/h^2) - f'(0) = 4s_E(L/h^2)\,. \tag{24}$$

Each position of the disc surface does not generally turn back to the same one after one revolution, but it can approximately turn back to the same one after two or more revolutions. In this case, it is only necessary to substitute L into $n*L$ (n: revolution number).

2.3.2. Ratio *Ra* of the Intersection between the Actual and Approximate Areas

In this section, we investigate whether Eq. (23) holds when the three sensors are assumed to move on a circumference of about the same circle. Here, each sensor is assumed to be a capacity sensor whose measuring value can be given by an average value of its sensing area. Especially, we apply the fact that the zero-difference of the sensors in a roundness measuring machine can be autonomously decided due to the circumferential measurement [18, 19].

The three sensors can not scan on the circumference of the same circle as is mentioned in section 3.1, and the relation between the scanning radius r_B of the sensors B and the scanning radius r_{AC} of the sensors A, C is always $r_{AC} > r_B$. Therefore, it is an essential problem how accurately Eqs. (8) - (10) can describe the behavior of the three sensors in our system.

Now, assume the following:

[1] Measuring area of each sensor is a circle with radius r_{sen}.

[2] Middle sensor B rotates around the circumference of a circle with radius r_B from a rotational center point O.

[3] Sensors A and C are arranged so that a line that connects the rotational center point O and the middle sensor B is perpendicular to a line that connects the sensor A and C.

When the sensor B moves on the circumference of a circle with radius r_B, the sensors A and C always move on one with radius $r_{AC}=[r_B^2+(h/2)^2]^{1/2}$, as is shown by Fig. 4.

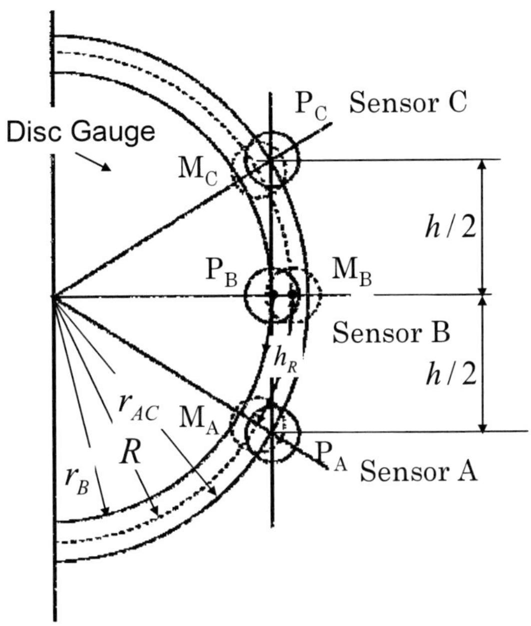

Figure 4. Positions P_A, P_B, P_C are points through which sensors really pass, and positions M_A, M_B, M_C are points through which sensors are assumed to pass in algorithm for deciding zero-difference s_E.

If the three sensors A, B and C are supposed to move on the circumference of a circle with radius $r_{ABC}=(r_B+r_{AC})/2$, the difference between the position at which they are supposed to move for deriving our algorithm and the position really measured by them is described by

$$\Delta r = \Delta r\left(h, r_B\right) = \frac{r_{AC} - r_B}{2} = \frac{\sqrt{\left(h/2\right)^2 + r_B^{\;2}} - r_B}{2}\left(\cong \frac{h^2}{16 r_B}\right) \tag{25}$$

Then, the ratio *Ra* of the common domain between the measuring area when the central position of the sensor A exists at M_A and at P_A as shown in Fig. 5 can be obtained by

$$Ra = Ra\left(r_{sen}, h, r_B\right)$$
$$= \frac{1}{\pi}\left(2\arctan\left(\frac{2r_{sen}}{\Delta r\left(h,r_B\right)}\sqrt{1-\left(\frac{\Delta r\left(h,r_B\right)}{2r_{sen}}\right)^2}\right) - \frac{\Delta r\left(h,r_B\right)}{r_{sen}}\sqrt{1-\left(\frac{\Delta r\left(h,r_B\right)}{2r_{sen}}\right)^2}\right) \text{[12]} \tag{26}$$

Note that points M_A and P_A denote the approximate measurement point and the real measurement point, respectively.

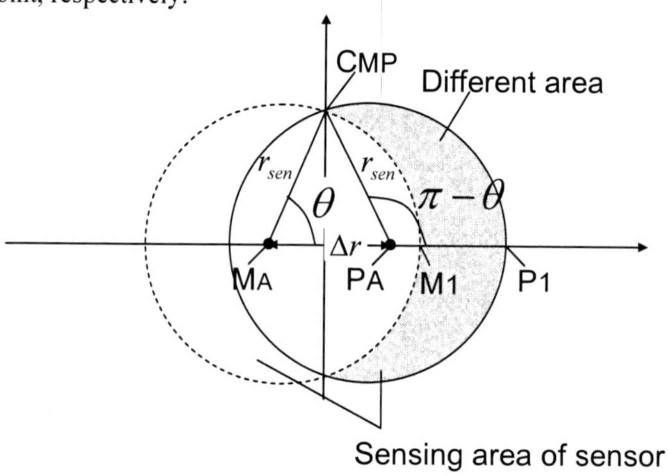

Figure 5. Decision of ratio *Ra* of common area to sensing area of sensors.

The ratio *Ra* for the sensor B and C can be obtained in same way.

The distance h_r along the arc line between the positions M_B and M_A (or M_C) as is shown in Fig.4 can be described by

$$h_r = r_{ABC} \times \arctan\left(\frac{h}{2r_B}\right). \tag{27}$$

Here, replacing the variable h in Eqs. (8) - (10) with the variable h_r described by Eq. (27), the equations (8) - (10) become

$$\mu_A(x) = C_A\left[f(x+h) - g(x) - e_P(x)\cdot h_r\right] + s_A, \tag{28}$$

$$\mu_B(x) = C_B\left[f(x+h/2) - g(x) - e_P(x)\cdot h_r/2\right] + s_B, \tag{29}$$

$$\mu_C(x) = C_C\left[f(x) - g(x)\right] + s_C. \tag{30}$$

Equations (28)-(30) denote the observation equations which the sensors A, B and C move on the circumference of the same circle with radius r_{ABC}.

2.4. Uncertainty Analysis

The uncertainty components are assumed as follows:

[1] The uncertainty components in case of the calibration process
(a) the linearity of the sensor; (b) the sensitivity of the sensor; (c) the zero-value of the sensor; (d) the adjacent sensor interval; (e) the repeatability of the measurement; (f) the fluctuation of the measurement position; (g) the different measurement positions, (h) the temperature change; and (i) the motion errors of the rotating table.

[2] The uncertainty components in case of the measuring process
(a) the linearity of the sensor; (b) the sensitivity of the sensor; (c) the zero-value of the sensor, (d) the adjacent sensor interval, (e) the temperature change, (f) the repeatability of the measurement; and (g) the fluctuation of the measurement position.

In this chapter, the uncertainty of the measurement result is assumed to be primarily caused by two specific uncertainty components. One of these is the repeatability of the sensor measurements; this component includes both the repeatability of the calibration and the repeatability of the measurement [(1) (e) and (2) (f)]. The other is the difference between the measurement positions of the sensors [(1) (f)].

The linearity and the sensitivity of the sensor are assumed to be able to determine exactly. And the temperature change and the motion error of the rotating table are assumed to be negligible when considering the feasibility of a rotating table. The other components will be studied experimentally.

Some of the properties of these uncertainty components are described in the next section.

2.4.1. Assumption for the Evaluation of Uncertainty

In this paper, the uncertainty analysis with the uncertainty components that cause the uncertainty of the measured shape of a target object is mainly carried out by considering their deviations.

We make the following assumptions:

(1) The uncertainty components associated with the zero-value (s_A, s_B, s_C)

Their deviations and standard uncertainties are assumed to be denoted by $(\Delta s_A, \Delta s_B, \Delta s_C)$, and $(\delta s_A, \delta s_B, \delta s_C)$, respectively.

Then, each average value of Δs_j ($j = A$, B, and C) is zero. The mutual correlation between the deviation of one sensor and that of other sensor is assumed to be zero and that between this uncertainty component and other one are assumed to be zero.

Furthermore, the following relations are assumed.

$$\delta_{\Delta s_k} = \delta_{\Delta s} \quad (k = A,\ B, \text{ and } C) \tag{31}$$

and the maximum value s_{max} of each zero-value (s_A, s_B, s_C) is assumed to be defined by

$$\operatorname*{Max}_{j=A,B,C} s_j = s_{\max}\,. \tag{32}$$

(2) The uncertainty components associated with the repeatability of measurement by the three sensors: $(\omega_A(l), \omega_B(l)$, and $\omega_C(l))$ and $(\omega_A(x), \omega_B(x)$, and $\omega_C(x))$

The average deviations and covariance values of the sensor output for repeated measurements for any sensor position are zero.

The standard uncertainties of the sensor output for repeated measurements are $\left(\delta_{\omega_A(l)}, \delta_{\omega_B(l)}, \delta_{\omega_C(l)}\right)$, and $\left(\delta_{\omega_A(x)}, \delta_{\omega_B(x)}, \delta_{\omega_C(x)}\right)$

These terms are the same with respect to the three sensors A, B and C and any position, i.e.,

$$\delta_{\omega_k(l)} = \delta_{\omega_k(x)} = \delta_{\omega} \quad (k = A, B, C) \tag{33}$$

and also the mutual correlation between this uncertainty component and other one is assumed to be zero.

Furthermore, the constant number ω_{max} is assumed to be defined by

$$\operatorname*{Max}_{\substack{j=A,B,C \\ 0 \le x \le X}} \left[\operatorname{abs}\left(\omega_j(x)\right)\right] = \omega_{\max} \tag{34}$$

(3) The uncertainty components caused by the different measurement positions: $(\mu^P_A(l), \mu^P_B(l), \mu^P_C(l))$

The deviations of these components are assumed to be $(\Delta\mu^P_A(l), \Delta\mu^P_B(l), \Delta\mu^P_C(l))$, respectively. The three sensors cannot scan on the same circumference of the circle with radius r from the rotational center point O because the flatness of disc gauge is not absolute.

The uncertainty caused by the different measurement positions is assumed as follows:

(a) The average deviations and covariance values of the height caused by the different measurement positions are constant with respect to the variable l, i.e.,

$$E\left[\Delta^P{}_A(l)\right]=E\left[\Delta^P{}_C(l)\right]=\Delta_{AC}{}^P,\; E\left[\Delta^P{}_B(l)\right]=\Delta_B{}^P, \tag{35}$$

$$E\left[\left(\Delta^P{}_A(l)\text{-}\Delta_{AC}{}^P\right)\left(\Delta^P{}_B(l)\text{-}\Delta_B{}^P\right)\right]=E\left[\left(\Delta^P{}_B(l)\text{-}\Delta_B{}^P\right)\left(\Delta^P{}_C(l)\text{-}\Delta_{AC}{}^P\right)\right]=0\,, \tag{36}$$

and

$$E\left[\left(\Delta^P{}_A(l)\text{-}\Delta_{AC}{}^P\right)\left(\Delta^P{}_C(l)\text{-}\Delta_{AC}{}^P)\right)\right]=\delta_{\Delta^P{}_{AC}}\,. \tag{37}$$

(b) The standard uncertainties of the height at the different measurement positions are $\left(\delta_{\Delta^P{}_A(l)},\,\delta_{\Delta^P{}_B(l)},\,\delta_{\Delta^P{}_C(l)}\right)$.

These terms are the same with respect to the three sensors A, B and C and any position (r, l), i.e.,

$$\delta_{\Delta^P{}_A(l)}=\delta_{\Delta^P{}_B(l)}=\delta_{\Delta^P{}_C(l)}=\delta_{\Delta P}\,. \tag{38}$$

Here,

$$\delta^2{}_{\Delta^P{}_j(l)}=E\left[\left(\Delta^P{}_j(l)\text{-}E\left[\Delta^P{}_j(l)\right]\right)\right]^2+E\left[\Delta^P{}_j(l)\right]^2\;(j\text{=A,B, and C}). \tag{39}$$

For simplicity, the following relation is assumed:

$$\delta_{\Delta^P{}_{AC}}=\delta^2{}_{\Delta P} \tag{40}$$

.

Here, the relation $\delta_{\Delta^P{}_{AC}}\le\delta^2{}_{\Delta P}$ is always approved from equation Eq. (37) and so $\delta_{\Delta^P{}_{AC}}$ is assumed to be maximum value for uncertainty analysis.

Furthermore, the mutual correlation between this uncertainty component and other one is assumed to be zero.

Here, the ratio Ra of the common domain between the sensor area that is approximated to drive the zero-difference and the actual sensor area is dependent on the radius r as is shown by Eq. (26).

Finally, the deviation of the measurement value $\Delta\mu_{kp}(l)$ ($k = A, B$ and C) of each sensor at position l on the circumference of the circle with radius r from the rotational center point O is described by the sum of the uncertainty component associated with the repeatability of measurement by the three sensors and the uncertainty component caused by the difference in the measurement positions, and can be obtained by the following equation:

$$\begin{aligned}\Delta_{\mu kp}(r,l)&=Ra(r_{sen},h,r)\Delta_{\mu k}(r,l)+\left(1-Ra(r_{sen},h,r)\right)\left(\Delta_{\mu k}(r,l)+\Delta^P{}_k(r,l)\right)\\&=\Delta_{\mu k}(r,l)+\left(1-Ra(r_{sen},h,r)\right)\Delta^P{}_k(r,l)\quad(k=\text{A, B, and C})\end{aligned} \tag{41}$$

where

$\Delta_{\mu k}(r,l),\Delta^{P}{}_{k}(r,l)$ $(k=A,B,C)$ indicate
$\Delta_{\mu A}(r,l),\Delta^{P}{}_{A}(r,l),\Delta_{\mu B}(r,l),\Delta^{P}{}_{B}(r,l),\Delta_{\mu C}(r,l),\text{and}\,\Delta^{P}{}_{C}(r,l)$.

2.4.2. Derivation of the Evaluation Equation of the Measured Straightness of the Object Surface

Based on the assumptions as mentioned in section 4.1, we investigate the expanded uncertainty U of the measured surface $f(x)$ of the object surface in this section.

In a first step, we consider the average deviations and standard uncertainties of the output values $\mu_A(l)$, $\mu_B(l)$, $\mu_C(l)$ which are measured across the circumference of a circle with radius r when calibrating the zero-difference.

The deviations of the output values of sensors : $\mu_A(l)$, $\mu_B(l)$, and $\mu_C(l)$ are assumed to be described by

$$\Delta\mu_j(l) = \Delta s_j + \omega_j(l) \tag{42}$$

with the zero-value and repeatability components Δs_j, $\omega_j(l)$ ($j=A$, B, and C). Therefore, the standard uncertainties $\delta\mu_A(l)$, $\delta\mu_B(l)$, $\delta\mu_C(l)$ of the measured output values of sensors $\mu_A(l)$, $\mu_B(l)$, and $\mu_C(l)$ are can be expressed as

$$\delta_{\Delta\mu_j(l)} = \left[\delta^2{}_{\Delta s} + \delta^2{}_{\omega}\right]^{1/2} (j=A,\ B,\ \text{and}\ C). \tag{43}$$

by the standard uncertainties of zero-value and repeatability components: $\delta_{\Delta s_j}$, δ_{ω_j}.

As well, the average value and standard uncertainty with measured values $\mu_A(x)$, $\mu_B(x)$, $\mu_C(x)$ along the straight line of the measuring target can be described in the same way.

As is shown by Eq. (41), the deviation $\Delta_{\mu kp}(l)$ of the measurement value of each sensor at position l on the circumference of the circle with radius R is assumed to be expressed by the sum of the deviation of uncertainty caused by the difference measurement position and the deviation of measurement uncertainty including other uncertainty components as the following equation

$$\delta_{\Delta_{\mu jp}(r,l)} = \left[\delta^2{}_{\Delta s} + \delta^2{}_{\omega} + \left(1 - Ra\left(r_{sen}, h, r\right)\right)^2 \delta^2{}_{\Delta^P{}_j(r,l)}\right]^{1/2} \tag{44}$$

($j=A$, B, and C)

Under the assumption mentioned above, we carry out the evaluation of the uncertainty of each component used to determine the shape $f(x)$ of the object surface [21].

(1) The standard uncertainty $\delta_{\Delta M(x)}$ of $\Delta M(x)$

First, the deviation of the $M_r(l)$ can be expressed by Eqs. (20), and (41) as

$$\Delta M_r(l) = \left(\frac{2}{h_r}\right)^2 \left[\frac{C_A \Delta\mu_A(l)}{C^2_A} - 2\frac{C_B \Delta\mu_B(l)}{C^2_B} + \frac{C_C \Delta\mu_C(l)}{C^2_C} + (1-Ra)\left(\frac{\Delta_A{}^P(l)}{C_A} + \frac{\Delta_C{}^P(l)}{C_C} - 2\frac{\Delta_B{}^P(l)}{C_B}\right)\right] \quad (45)$$

By this equation, the average of the $\Delta M_r(l)$ can be given by

$$E[\Delta M_r(l)] = \left(\frac{2}{h_r}\right)^2 (1-Ra)\left(\left(\frac{1}{C_A} + \frac{1}{C_C}\right)\Delta_{AC}{}^P - 2\frac{\Delta_B{}^P}{C_B}\right), \quad (46)$$

Because $E\left[\Delta\mu_j(l)\right] = 0$ $(j = A, B, \text{and } C)$ are approved.

Therefore, the standard uncertainty $\delta_{\Delta M(x)}$ of $\Delta M(x)$ is described by

$$\begin{aligned}
\delta_{\Delta M_r(l)} &= \left(E\left[\Delta M_r(l) - E\left[\Delta M_r(l)\right]\right]^2 + E\left[\Delta M_r(l)\right]^2\right)^{1/2} \\
&= \left(\frac{2}{h_r}\right)^2 \left(\frac{1}{C^2_A}\delta^2{}_{\Delta\mu_A(l)} + 4\frac{1}{C^2_B}\delta^2{}_{\Delta\mu_B(l)} + \frac{1}{C^2_C}\delta^2{}_{\Delta\mu_C(l)} + (1-Ra(r))^2 u_{Mp}\delta^2{}_{\Delta P}\right)^{1/2} \\
&= \left(\frac{2}{h_r}\right)^2 \left(u_{Ms\omega}\delta^2{}_{\Delta s} + u_{Ms\omega}\delta^2{}_{\omega} + (1-Ra(r))^2 u_{Mp}\delta^2{}_{\Delta P}\right)^{1/2}
\end{aligned} \quad (47)$$

where

$$u_{Ms\omega} = \frac{1}{C^2_A} + \frac{4}{C^2_B} + \frac{1}{C^2_C}, \quad (47)$$

$$u_{Mp} = \left(\frac{1}{C_A} + \frac{1}{C_C}\right)^2 + \frac{4}{C^2_B}. \quad (48)$$

(2) The standard uncertainty $\delta s_{E(r)}$ of $\Delta s_{E(r)}$

The average deviation $E[\Delta s_{E(r)}]$ is determined from Eq. (23) and (45) by

$$E[\Delta s_E] = \frac{h_r^2}{4L}\oint_{0\le\alpha\le L} E\left[\Delta M_r(\alpha)\right]d\alpha = (1-Ra)\left(\frac{\Delta_{AC}{}^P}{C_A} + \frac{\Delta_{AC}{}^P}{C_C} - 2\frac{\Delta_B{}^P}{C_B}\right) \quad (50)$$

and standard deviation of $\delta s_{E(r)}$ can be expressed as

$$\begin{aligned}
\delta_{\Delta s_E(r)}{}^2 &= E\left[\Delta s_E(r) - E\left[\Delta s_E(r)\right]\right]^2 + E\left[\Delta s_E(r)\right]^2 \\
&= \frac{h_r{}^4}{16(2\pi r)^2} E\left[\oint_{0\le\alpha\le 2\pi r}\left(\Delta M_r(r,\alpha) - E\left[\Delta M_r(r,\alpha)\right]\right) d\alpha\right]^2 \\
&= \frac{s^2}{(2\pi r)^2} E\left[\sum_{k=1}^{n_r}\left(\Delta\mu_A(r,\, l_k) + \Delta\mu_C(r,\, l_k) - 2\Delta\mu_B(r,\, l_k) + \Delta_{ABC}{}^P(l_k)\right)\right]^2 \\
&\qquad\qquad (n_r s = 2\pi r) \\
&= \frac{u_{Ms\omega}}{n_r{}^2}\left(\left(n_r + 2_{n_r}C_2\right)\delta^2{}_s + n_r\delta^2{}_\omega\right) + \left(1 - Ra(r)\right)^2 u_{Mp}\delta^2{}_{\Delta P} \\
&= \frac{u_{Ms\omega}}{n_r}\delta^2{}_\omega + u_{Ms\omega}\delta^2{}_s + \left(1 - Ra(r)\right)^2 u_{Mp}\delta^2{}_{\Delta P}
\end{aligned} \tag{49}$$

where

$$\Delta_{ABC}{}^P(l) = \left(\frac{\Delta_A{}^P(l) - \Delta_{AC}{}^P}{C_A} + \frac{\Delta_C{}^P(l) - \Delta_{AC}{}^P}{C_C} - 2\frac{\Delta_B{}^P(l) - \Delta_B{}^P}{C_B}\right)\Bigg] \tag{50}$$

This relation can be approved under the assumptions (1) ~ (5) stated in Section 4.1.

(3) The standard uncertainty $\delta_{\iint_{0\le\alpha\le s,\,0\le s\le x}\Delta M(\alpha)d\alpha ds}$ generated when measuring using three sensors

The deviation $\Delta M(x)$ of $M(x)$, which is calculated by measuring along the straight line of an object surface, is described by

$$\Delta M(x) = \left(\frac{2}{h}\right)^2\left(\Delta\mu_A(x) + \Delta\mu_C(x) - 2\Delta\mu_B(x)\right) \tag{53}$$

Therefore, the average value of $\Delta M(x)$ is zero for each x, and the standard uncertainty $\delta_{\Delta M(x)}$ of $\Delta M(x)$, which is calculated by measuring along the straight line of an object surface, is described by

$$\delta_{\Delta M(x)} = \left(\frac{2}{h}\right)^2\left(u_{Ms\omega}\delta^2{}_{\Delta s} + u_{Ms\omega}\delta^2{}_\omega\right)^{1/2}. \tag{51}$$

Next, the standard uncertainty of the double linear integral of $\Delta M(x)$ along the straight line of the object surface is also given by

$$\delta_{\iint_{0\le\alpha\le s,\,0\le s\le x}\Delta M(\alpha)d\alpha ds} = \frac{4x^2}{h^2 n_x^{\;2}}\left[\sum_{k=1}^{n_x}(n_x-k+1)^2 u_{Ms\omega}\delta^2_{\;\omega} + \left(\sum_{k=1}^{n_x}(n_x-k+1)\right)^2 u_{Ms\omega}\delta^2_{\;\Delta s}\right]^{1/2} \tag{52}$$

where (a) x is the measurement length, (b) n_x is the sampling number of the measurement, and (c) $x=n_x s$, and $x_{k+1}-x_k=s$.

The double linear integral $\iint_{0\le\alpha\le s,\,0\le s\le x}\Delta M(\alpha)d\alpha ds$ of $\Delta M(x)$ along the straight line of the object surface is described from Eq. (13) by

$$\begin{aligned}\iint_{0\le\alpha\le s,\,0\le s\le x}\Delta M(\alpha)d\alpha ds &\cong \sum_{m=0}^{n_x} s\sum_{k=0}^{m}\Delta M(x_k)s\\ &= \frac{4}{h^2}s^2\sum_{m=0}^{n_x}\sum_{k=0}^{m}\left[\frac{C_A\Delta\mu_A(x_k)-\Delta C_A\mu_A(x_k)}{C^2_{\;A}} - 2\frac{C_B\Delta\mu_B(x_k)-\Delta C_B\mu_B(x_k)}{C^2_{\;B}} + \frac{C_C\Delta\mu_C(x_k)-\Delta C_C\mu_C(x_k)}{C^2_{\;C}}\right]\end{aligned} \tag{53}$$

Therefore, the variance of $\iint_{0\le\alpha\le s,\,0\le s\le x}\Delta M(\alpha)d\alpha ds$ can be described by

$$\begin{aligned}\delta^2_{\iint_{0\le\alpha\le s,\,0\le s\le x}\Delta M(\alpha)d\alpha ds} &= E\left[\left(\iint_{0\le\alpha\le s,\,0\le s\le x}\Delta M(\alpha)d\alpha ds\right)^2\right] \cong E\left[\left(\sum_{m=0}^{n_x} s\sum_{k=0}^{m}\Delta M(x_k)s\right)^2\right]\\ &= \frac{16s^4}{h^4}E\left[\sum_{m=1}^{n_x}\sum_{k=1}^{m}\left(\frac{C_A\Delta\mu_A(x_k)}{C^2_{\;A}} - 2\frac{C_B\Delta\mu_B(x_k)}{C^2_{\;B}} + \frac{C_C\Delta\mu_C(x_k)}{C^2_{\;C}}\right)^2\right]\end{aligned} \tag{54}$$

Here, the following relation can be approved.

$$\begin{aligned}&\sum_{m=1}^{n_x}\sum_{k=1}^{m}\Delta\mu_s(x_k)\\ &= n_x\Delta\mu_s(x_1) + (n_x-1)\Delta\mu_s(x_2) + \cdots + (n_x-p-1)\Delta\mu_s(x_p) + \cdots + \Delta\mu_s(x_{n_x})\\ &= \sum_{k=1}^{n_x}(n_x-k+1)\Delta\mu_s(x_k)\end{aligned} \tag{55}$$

where

$$\Delta\mu_s(x_k) = \frac{\Delta\mu_A(x_k)}{C_A} - 2\frac{\Delta\mu_B(x_k)}{C_B} + \frac{\Delta\mu_C(x_k)}{C_C}. \tag{56}$$

Therefore, Eq. (55) is also described by

$$\delta^2_{\iint_{0\le\alpha\le s,0\le s\le x}\Delta M(\alpha)d\alpha ds} = \frac{16s^4}{h^4}E\left[\sum_{k=1}^{n_x}(n_x-k+1)\Delta\mu_s(x_k)\right]^2 = \frac{16s^4}{h^4}\left[\sum_{k=1}^{n_x}(n_x-k+1)^2 u_{Ms\omega}\delta^2_{\omega}+\left(\sum_{k=1}^{n_x}(n_x-k+1)\right)^2 u_{Ms\omega}\delta^2_{\Delta s}\right]^2, \tag{57}$$

where $x=n_x s$ and $x_{k+1}-x_k=s$.

Finally, the standard deviation of $\iint_{0\le\alpha\le s,0\le s\le x}\Delta M(\alpha)d\alpha ds$ can be determined by

$$\delta_{\iint_{0\le\alpha\le s,0\le s\le x}\Delta M(\alpha)d\alpha ds} = \frac{4x^2}{h^2 n_x^2}\left[\sum_{k=1}^{n_x}(n_x-k+1)^2 u_{Ms\omega}\delta^2_{\omega}+\left(\sum_{k=1}^{n_x}(n_x-k+1)\right)^2 u_{Ms\omega}\delta^2_{\Delta s}\right]^{1/2}. \tag{58}$$

2.4.3. Expanded Uncertainty *U* of the Measurement Shape *f(x)* of an Object Surface

In this section, we mainly investigate the relationship between the expanded uncertainty *U* of the measured shape of an object target and the uncertainty components used to determine the object surface obtained in the previous section.

The average deviation $E[\Delta f(x)]$ of the measured shape $f(x)$ of an object surface is described by

$$E[\Delta f(x)] = \iint_{0\le\alpha\le s,0\le s\le x}E[\Delta M(\alpha)]d\alpha ds - 2\left(\frac{x}{h}\right)^2 E[\Delta s_E] = -2(1-Ra)\left(\frac{\Delta_{AC}^{\ \ P}}{C_A}+\frac{\Delta_{AC}^{\ \ P}}{C_C}-2\frac{\Delta_B^{\ P}}{C_B}\right)\left(\frac{x}{h}\right)^2 \tag{59}$$

from Eqs. (40) and (47), and the combined standard uncertainty $\delta\Delta f(x)$ of the measured shape $f(x)$ of the object surface is also expressed by Eqs. (12) and (52) as

$$\delta_{\Delta f(x)} = \sqrt{\delta^2_{\iint_{0\le\alpha\le s,\,0\le s\le x}\Delta M(\alpha)d\alpha ds} + \left(2\left(\frac{x}{h}\right)^2\right)^2 \delta^2_{\Delta s_E(r)}}$$
$$= \frac{2x^2}{h^2}\left[\frac{2}{n_x^{\ 4}}\left(\sum\nolimits_{k=1}^{n_x}(n_x-k+1)^2 u_{Ms\omega}\delta^2_{\ \omega} + \left(\sum\nolimits_{k=1}^{n_x}(n_x-k+1)\right)^2 u_{Ms\omega}\delta^2_{\ \Delta s}\right)\right.$$
$$\left. + \frac{u_{Ms\omega}}{n_r}\delta^2_{\ \omega} + u_{Ms\omega}\delta^2_{\ s} + (1-Ra(r))^2 u_{Mp}\delta^2_{\ \Delta P}\right]^{1/2} \tag{60}$$

Eq. (60) can be rewritten by

$$\delta_{\Delta f(x)} = \frac{2x^2}{h^2}\left[k_\omega\delta^2_{\ \omega} + k_s\delta^2_{\ s} + k_p\delta^2_{\ \Delta P}\right]^{1/2}, \tag{61}$$

where

$$k_\omega = \frac{4}{n_x^{\ 4}}\sum\nolimits_{k=1}^{n_x}(n_x-k+1)^2 u_{Ms\omega} + \frac{u_{Ms\omega}}{n_r}, \tag{62}$$

$$k_s = \frac{4}{n_x^{\ 4}}\left(\sum\nolimits_{k=1}^{n_x}(n_x-k+1)\right)^2 u_{Ms\omega} + u_{Ms\omega}, \tag{63}$$

and

$$k_p = (1-Ra)^2 u_{Mp}. \tag{64}$$

Therefore, the expanded uncertainty $U=U(x)$ of the measured shape $f(x)$ of the object surface is described by

$$U = 2\delta_{\Delta f(x)}$$
$$= \frac{4x^2}{h^2}\left[k_\omega\delta^2_{\ \omega} + k_s\delta^2_{\ s} + k_p\delta^2_{\ \Delta P}\right]^{1/2} \quad (k=2). \tag{65}$$

2.4.4. Evaluation of Expanded Uncertainty *U*

In this section, we investigate the expanded uncertainty U of the measured shape $f(x)$. The coefficient k_w of standard uncertainty δ_w can be evaluated by

$$
\begin{aligned}
k_{\omega} &= \frac{4}{n_x^{\ 4}}\sum_{k=1}^{n_x}\left(n_x - k + 1\right)^2 u_{Ms\omega} + \frac{u_{Ms\omega}}{n_r} \\
&= \left(\frac{4}{n_x^{\ 4}}\left(\frac{2n_x^{\ 3} + 3n_x^{\ 2} + n_x}{6}\right) + \frac{1}{n_r}\right)u_{Ms\omega} \\
&= 4\left(\frac{1}{3n_x} + \frac{1}{2n_x^{\ 2}} + \frac{1}{6n_x^{\ 3}} + \frac{1}{4n_r}\right)u_{Ms\omega}
\end{aligned}
\tag{66}
$$

Because of

$$
\sum_{k=1}^{n_x}\left(n_x - k + 1\right)^2 = \sum_{k=1}^{n_x} k^2 = \frac{2n_x^{\ 3} + 3n_x^{\ 2} + n_x}{6}
\tag{67}
$$

Also, the coefficient k_s of standard uncertainty δ_s can be evaluated by

$$
\begin{aligned}
k_s &= \frac{4}{n_x^{\ 4}}\left(\sum_{k=1}^{n_x}\left(n_x - k + 1\right)\right)^2 u_{Ms\omega} + u_{Ms\omega} \\
&= 4\left(\frac{1}{4}\left(1 + \frac{1}{n_x}\right)^2 + \frac{1}{4}\right)u_{Ms\omega}
\end{aligned}
\tag{68}
$$

Because of

$$
\frac{1}{n_x^{\ 4}}\left(\sum_{k=1}^{n_x}\left(n_x - k + 1\right)\right)^2 = \frac{1}{n_x^{\ 2}}\left(\sum_{k=1}^{n_x} k\right)^2 = \frac{\left(n_x + 1\right)^2}{4n_x^{\ 2}},
\tag{69}
$$

Therefore, the expanded uncertainty $U=U(x)$ of the measured shape $f(x)$ of the object surface is rewritten by

$$
\begin{aligned}
U = \frac{4x^2}{h^2}\Bigg[&4\left(\frac{1}{C^2_{\ A}} + \frac{4}{C^2_{\ B}} + \frac{1}{C^2_{\ C}}\right)\left(\left(\frac{1}{3n_x} + \frac{1}{2n_x^{\ 2}} + \frac{1}{6n_x^{\ 3}} + \frac{1}{4n_r}\right)\delta^2_{\ \omega} + \left(\frac{1}{4}\left(1 + \frac{1}{n_x}\right)^2 + \frac{1}{4}\right)\delta^2_{\ s}\right) \\
&+ \left(1 - Ra\right)^2\left(\left(\frac{1}{C_A} + \frac{1}{C_C}\right)^2 + \frac{4}{C_B}\right)\delta^2_{\ \Delta P}\Bigg]^{1/2} \quad (k = 2)
\end{aligned}
\tag{70}
$$

If we assume large sampling number of the measurement n_r and n_x, the above Eq. (70) can be described by

$$U = \frac{4x^2}{h^2}\left[\frac{1}{2}\left(\frac{1}{C^2{}_A}+\frac{4}{C^2{}_B}+\frac{1}{C^2{}_C}\right)\delta^2{}_s + (1-Ra)^2\left(\left(\frac{1}{C_A}+\frac{1}{C_C}\right)^2+\frac{4}{C_B}\right)\delta^2{}_{\Delta P}\right]^{1/2} . \quad (k=2) \tag{71}$$

As is easily understood from Eq. (70), the uncertainty of the measured shape $f(x)$ becomes large in proportion to the value $(x/h)^2$ which is described by the measuring length x and the adjacent sensor interval $h/2$. The influence of random uncertainty components considered in this study on it can be reduced in proportion to the square root of the sampling number of the measurement when measuring area ($0<=x<=X$) is fixed. Furthermore, the influence of the systematic error components δ_s and $\delta_{\Delta P}$ remains in spite of the largeness of the sampling number of the measurement.

Remark

The result which can be obtained in this section is satisfied under the assumption that the measuring environment is temperature controlled one and the measurement time will be short (so, the other uncertainty factors such as the drift are assumed to be negligible).

2.4.5. Numerical Simulation and its Result

In this section, the expanded uncertainty U of the shape $f(x)$ of the object surface obtained by our method is evaluated under the adequate and practical assumptions. Here, we mainly investigate the influence of the uncertainty components associated with the repeatability of the sensor measurements, the difference between the measurement positions of the sensors, and the zero-value on the measured uncertainty.

The expanded uncertainty U of the straightness $f(x, r)$ of the object surface obtained by our method is evaluated based on the assumption of adequate and practical variables.

2.4.5.1. Suppositions for the Numerical Simulation

(1) The displacement sensor and its properties

The sensors are capacitive-type sensors whose sensing elements are circular and have radius r_{sen} of 1.0 mm. The corresponding effective measurement areas are set to be

$$r_{effective} = 1.7\ (\text{mm}), \tag{72}$$

because the effective area is about *5/3* times as large as the sensing area of the capacitive sensor [22].

The standard uncertainty of the sensor output for repeated measurements is assumed to be obtained from the uniform distribution of ±0.1 (μm) , and standard uncertainty is set to be

$$\delta_{\omega} = \frac{0.1}{\sqrt{3}}\,(\mu\text{m}) \tag{73}$$

considering the products.

In the same way, the uncertainty component associated with the zero-value is also assumed to be obtained from the uniform distribution of ±0.1 (μm) , and standard uncertainty is set to be

$$\delta_{\Delta s} = \frac{0.1}{\sqrt{3}}\,(\mu\text{m}) \tag{74}$$

The sensitivity coefficients are assumed to be set to be

$$C_A = C_B = C_C = 1. \tag{75}$$

The adjacent sensor interval *h/2* is set to be

$$h/2 = 30\ (\text{mm}). \tag{76}$$

(2) The disc gauge

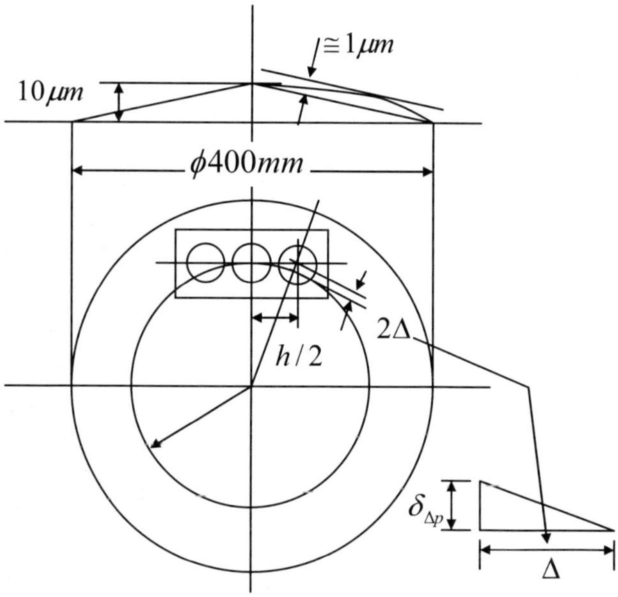

Figure 6. Disc gauge: (a) the diameter is 400 mm and (b) the peak height is 10 μm.

The disc gauge used to determine the zero-difference s_E is manufactured by a current CNC turning machine. The surface of the disc gauge can vary due to mechanical alignment errors. The surface is assumed to be conical in shape. For example, when the radius of the disc gauge is 200 mm, a peak height value of 10 μm is assumed and the magnitude of the low-frequency factor of its surface is approximately 1 μm, as shown in the upper side of Fig. 6.

For a disc gauge of this radius and peak height, the standard uncertainty of the straightness caused by the different measurement positions of the disc gauge is assumed to be

$$\delta_{\Delta P} = \left[\frac{\left(\sqrt{r^2 + (h/2)^2} - r \right)}{200 \times 2} \times 10 + 1/\cos\left(\frac{10}{200 \times 10^3} \right) \right] / \sqrt{3}\,(\mu\text{m}), \tag{77}$$

where r is the disc gauge radius. Eq. (77) shows how the difference in height between the center point of the actual sensor area and that of the approximate sensor area is used to derive the zero-difference, as shown in Fig. 6.

(3) The measuring length and sampling number

The measuring length l is restricted to the range 50 mm - 200 mm, and the sampling number is 500-200, respectively, i.e., one sampling per 1 mm; this is the same as in the calibration case.

Based on the above assumption, the evaluation equation of the measurement uncertainty of the measured shape of the object surface (70) can be expressed as

$$U = \frac{4\sqrt{2}x^2}{h^2} \left[\left(\frac{4}{n_x} + \frac{6}{n_x^{\ 2}} + \frac{2}{n_x^{\ 3}} + \frac{3}{n_r} \right) \delta^2_{\ \omega} + \frac{3}{2} \left(\left(1 + \frac{1}{n_x} \right)^2 + 1 \right) \delta^2_{\ s} + 4(1 - Ra)^2 \, \delta^2_{\ \Delta P} \right]^{1/2} \tag{78}$$

2.4.5.2. Numerical Simulations and Their Results

At first, the relations of the ratio R of intersection domain between the area measured really and the assumed area for deriving the zero-difference s_E are investigated. Fig. 7 shows the relation of the ratio R to the radius r_{sen} and the adjacent sensor interval *h/2*. In this case, the measuring radius r_B is set to be 200 mm. The ratio R of the intersection domain changes largely with the adjacent sensor interval *h/2*, and it increases in proportion to the radius r_{sen}.

Next, the relationship between the measurement uncertainty of the zero-differences and the size of the disc gauge used for calibration is analyzed and simulated numerically belie fly. It is proved that the use of a disc gauge of larger radius results in smaller uncertainty of straightness measurement using the evaluation equation of the measurement uncertainty of the measured shape of the object surface.

The relationship between the expanded uncertainty of measurement straightness and the radius of the disc gauge is obtained from equation (78) and shown in Fig. 8. The expanded uncertainties of measurement straightness U are =3.44, 2.55, 2.05, 1.79, and 1.69 (μm) when the radius of the disc gauge R are 60, 100, 140, 180, and 200 (mm) and measurement length x is 100 (mm)

The use of a larger disc gauge results in smaller measurement uncertainty, and the measurement uncertainty becomes larger when the length of the object increases.

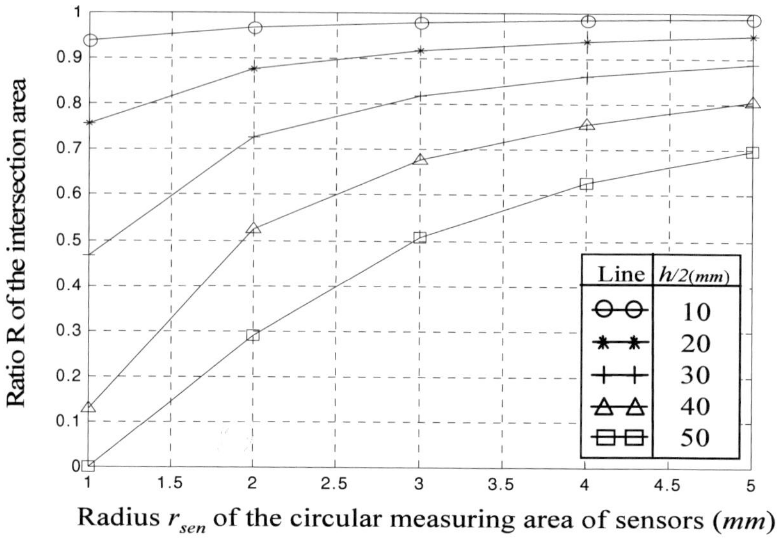

Figure 7. Relation of ratio R of intersection area to radius r_{sen} of circular measuring area of sensors when interval between sensors is 10, 20, 30, 40, 50 (mm) and radius r_b of sensor B is 200 (mm).

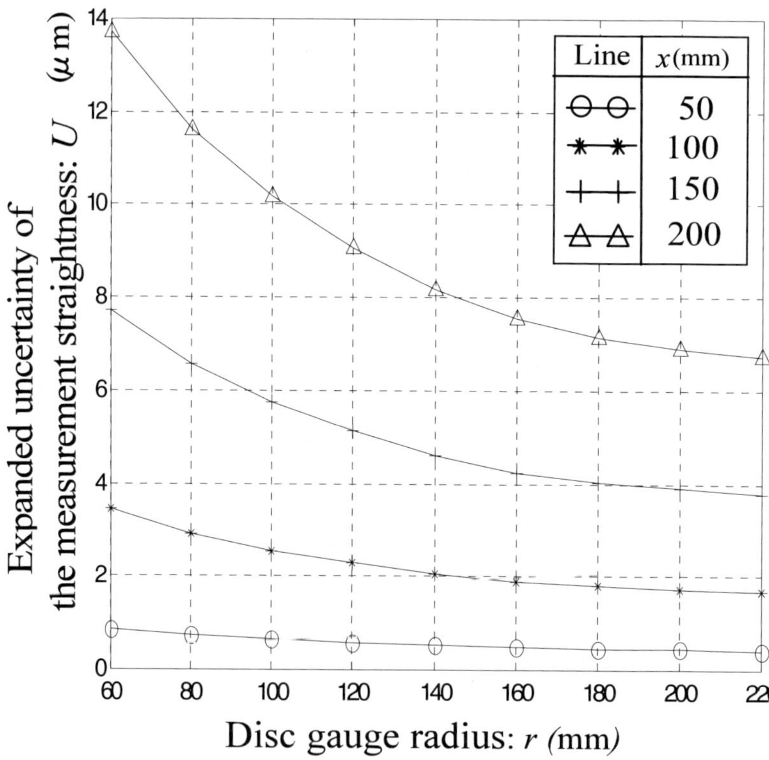

Figure 8. The relationship of the expanded uncertainty of measurement straightness U to the radius of the disc gauge R when measurement length x =50,100,150, and 200 (mm) and the adjacent sensor interval $h/2$ is 30 (mm).

2.5. CONCLUSIONS

In this chapter, an autonomous method for calibrating the zero-difference for the three-point method of measuring the surface straightness was explained with the uncertainty analysis in detail.

[1] The proposed autonomous calibration method has the following advantages:
   a. The calibration gauge is non-standard simple disc-shaped which can be made by the current CNC turning machine, etc.
   b. The zero-difference is only determined by two steps:
      Step1: Scan the disc gauge by rotation of three displacement sensors
      Step2: Calculate the zero difference by the proposed algorithm
   c. The method proposed in this paper is considered to be sufficiently practical and easy because the zero difference can be autonomously obtained.

[2] The relationship between the disc gauge radius and the measurement uncertainty was simulated numerically. The results of this study are as follows:
   a. The use of a disc gauge of larger radius results in smaller measurement uncertainty.
   b. If the measurement uncertainty of any object surface is given, the necessary size of the disc gauge radius can be easily derived using Fig.8.
   c. The expanded uncertainty U of measurement straightness of the object surface becomes less than 1.70 μm and less than 6.70 μm at measured lengths of 100 mm and 200 mm, respectively, when the radius of the disc gauge R is 200 mm, the sensor radius is 1.0 mm, the adjacent sensor interval is 30 mm, and the standard uncertainties associated with repeatability of the measurement and associated with the zero-value are assumed to be $0.1/\sqrt{3}$ μm.

[3] The method proposed in this paper is very practical for use with online applications for the measurement of the flatness of a next-generation 450-mm wafer for the following reasons:
   a. The mechanical construction is as simple as the conventional method that uses a single displacement sensor.
   b. Higher accuracy can be obtained using the three-point method.
   c. Calibration of the zero-difference for the three-point method is automatically carried out when measuring the flatness of the circumferential part of the wafer.

REFERENCES

[1] Farago FT and Curtis M A, *Handbook of dimensional measurement*.1994 (NY: Industrial Press Inc.).

[2] Thalmann R, 17 Straightness and Alignment in *Handbook of Optical Metrology*, 2009 (NY: CRC Press Inc.), 411-21.

[3] Tanaka H, Sato H. Study on the fundamental property with Sequential Two-point straightness method. *JSME*. 1982;48;436:1930-7(in Japanese).

[4] Tozawa K, Sato H, O-hori M. A new method for the measurement of the straightness of machine tools and machined work. *J Trans ASME Mech Design*. 1982;104:587–92.

[5] Tanaka H, Sato H. Extensive analysis and development of straightness measurement by sequential-two-points method. *Trans ASME J Eng Ind.* 1986;108:176–82

[6] Gao W, Kiyono S. High accuracy profile measurement of a machined surface by the combined method. *Measurement*. 1996;19(1):55–64.

[7] Gao W, Kiyono S. On-machine profile measurement of machined surface using the combined three-point method. *JSME Int J*. 1997;C 40:253–9.

[8] Tamura Y, Taira U, Sakamoto Y. Serial Multi-point Method for Surface Profile Measuring. *SICE*. 1989;25(5):509-16 (in Japanese).

[9] Kiyono S, Gao W, Kamada O. O-machine measurement of mirror profile. *J. Japan. Soc. Precision Eng.* 1992;58(2):247-52(in Japanese).

[10] Chen X, Kotani K, Takamasu K. Study on Spatial Frequency Domain 2-point method. *J. Japan.Soc. Precision Eng*. 2007;73(6):653-8(in Japanese).

[11] Kume T. Straightness measurement of X-band accelerator structure. In *Proceeding of Japan.Soc. Precision Eng Conference*. 2004 :1111-2(in Japanese).

[12] Yamaguchi J. Measurement of straight motion accuracy using the improved sequential three-point method. *J. Japan.Soc. Precision Eng.* 1993;59(5):773–8(in Japanese).

[13] Gao W, Yokoyama J, Kiyono K. Straightness Measurement of Cylinder by Multi-Probe Method. *Precision Eng.* 2002;28(3):279-89.

[14] Kume T. Straightness measurement of X-band accelerator structure. In *Proceeding of Japan.Soc. Precision Eng Conference* 2004:1111-2(in Japanese).

[15] Liu S, Watanabe K. Chen X. Takahashi S. Takamasu K. Profile measurement of a wide-area resist surface using a multi-ball cantilever system. *Precision Eng.* 2009;33:50-5.

[16] Elster C. Weingartner I. Coupled distance sensor systems for high-accuracy topography measurement. *Precision Eng*. 2006;30(1):32-8.

[17] Fujimoto I, Nishimura K, Pyoun Y. Autonomous Calibration Method of the Zero-difference without using a Standard Gauge for a Straightness-Measuring Machine. *Precision Eng*. 2011;35;1:153-63.

[18] Fujimoto I, Nishimura K, Takatsuji T, Pyun Y. A study on the relationship between measurement uncertainty and the size of the disc gauge used to calibrate a straightness measuring system. *Meas. Sci. Technol.* 2011 ; 22 ;125201.

[19] Fujimoto I, Nishimura K, Takatsuji T, Pyun Y. A Technique to Measure the Flatness of next-generation 450 mm wafers using a three-point method with an autonomous calibration function. *Precision Eng*. 2012 ;36 ; 270–280.

[20] I.Fujimoto,T. Takatsuji, K.Nishimura, and M.Y. Kim. Autonomous calibration method of the reference flat surface of an interferometer without using a standard flat surface. *Appl. Opt*. 2012;51;20;4754-4767.

[21] Internationl Organization for Standardization. *Guide to the expression of uncertainty in measurement (GUM)*. ISO,1995.

[22] KLA-Tencor Corp. Technical Data, Outline of Capacity Sensor. KLA-Tencor Japan Homepage *(http://www.KLA-Tencor.co.jp)*

In: Calibration Technology, Theories and Applications
Editors: Ikumatsu Fujimoto and Kunitoshi Nishimura
ISBN: 978-1-62618-808-2

Chapter 3

ON-SITE CALIBRATION METHOD OF A LARGE SCALE LASER DIGITIZER SYSTEM

Ikumatsu Fujimoto*
Okuwa Technical Research Center, Yamaguchi, Japan

ABSTRACT

In this chapter, the on-site calibration method of a two-dimensional large-scale laser digitizer with a cordless cursor is briefly proposed. The coordinate detecting scheme of this digitizer is based on the well-known triangulation method, in which two laser-rays are scanned by the rotating plane mirrors, reflected backward by the cursor, reflected again by the rotating mirrors, and detected by optical sensors. From angles in which the laser-rays reflected from cursor are detected by optical sensors, the position of the cursor can be determined. This method involves several problems on its calibration especially when it is applied to a large-scale digitizer. The coordinate calculation algorithm proposed in this chapter is determined by the system parameters such as the rotating center position of rotating mirror, the straight-line equation of the ray from the laser, etc. Especially, we propose the on-site calibrating method for determining the appropriate coordinate calculation algorithm, which gives the better coordinate accuracy.

Keywords: Digitizer, Coordinate error, Coordinate calculation algorithm, Control point, Optimization, On-site calibration, On-site insurance of measurement accuracy

3.1. INTRODUCTION

At first, the meaning of a term "on-site calibration" described in this chapter is to determine the appropriate coordinate calculation algorithm of a two-dimensional large-scale laser digitizer quickly by measuring the angle values at some control points whose positions are exactly determined beforehand at the measurement field.

* Corresponding Author: E-mail : fujimoto.ikumatsu@gmail.com

The use of digitizers has been spreading widely in many industrial areas such as construction, advertising, apparel and gardening because of the advantage of its automation design [1]. For coordinate detection, planar distributed devices such as magnetostrictive sheets have been extensively-used [2-3]. However, this detecting principle makes it difficult to apply to a portable device having an increased scale.

To solve such problems, we investigated the digitizer with optical method [4] and have developed a two-dimensional large-scale digitizer system using laser-ray with cordless cursor [5-7], which is the device to input quickly and freely the coordinates of large-scale graphical drawnings into a computer and it is really used in the area of manufacturing design, construction, surveying [8].

For example, about 2 m×6 m sized one is used to design a car body with full scale and about 2 m×3 m sized one is used to design glass products with full scale. High accuracy and easiness of manipulation are also required to such type of digitizer. Fig. 1 shows one of the typical one of systems. In this system, there are two optical units $U^{(1)}$ and $U^{(2)}$, each of which includes a laser, a rotating plane mirror, a half mirror and a sensor. The cordless cursor has a retro-reflective mirror and can move in the coordinate area of the digitizer. Both rotating plane mirrors rotate at about 2400 rpm, and the reflecting laser-ray from the mirrors scans the coordinate area.

One of the most difficult problems for large scale digitalization is whether it can be kept the measurement accuracy without losing easiness of handling and portability in practical situations or not. When we move the above large size digitizing device from one measurement place to another place, we have to take it to pieces and build it up again. It is impossible to assure the measuring accuracy without calibration, and to keep the surface flat, the temperature, and other measuring condition stationary.

We gave the solutions to such problems from both the architecture and the algorithm of the digitizing system to realize the easy on-site calibration with on-site insurance of measurement accuracy. An architectural solution is to use laser beam triangulation method. The use of the triangulation makes it possible for us to concentrate the equipments requiring geometrical accuracy into small number of small units. This will be essential for the compatibility of portability and accuracy. An algorithmic solution is to use the linearized algorithm which is invariant with alignment mismatch and mutual distance inaccuracy and so on. By the triangulation system and linearized algorithm, the on-site calibration with on-site error analysis can be performed very quickly.

The parameters of algorithm are detemined by measuring four control points whose coordinates are known beforehand, but the coordinate accuracy is largely different by the positions of four control points. Therefore, it is neccesary to study the method for determinig the optimal four control points in coordinate accuracy.

In this chapter, the on-site calibration method of a large scale laser digitizer with on-site insurance of measurement accuracy is proposed with error analysis and basic experiment. This chapter is organized as follows. The outline of laser digitizer system and its algorithm are simply given in Section 3.2. The error analysis is carried out with the description by error ellipse in Section 3.3. The optimal control points are detemined by a conventional optimazation method in Section 3.4. The experiment and its result are given in Section 3.5. Finally, a simple conclusion is outlined in Section 3.6 .

3.2. Coordinate Calculation Algorithm of a Laser Digitizer System for On-Site Calibration

3.2.1. System Configuration

Fig. 1 and Fig. 2 show a laser digitizer system with a cordless cursor. In this system, there are two optical units $U^{(1)}$ and $U^{(2)}$, each of which includes a laser, a rotating plane mirror installed in a motor, a polarized beam splitter and a sensor.

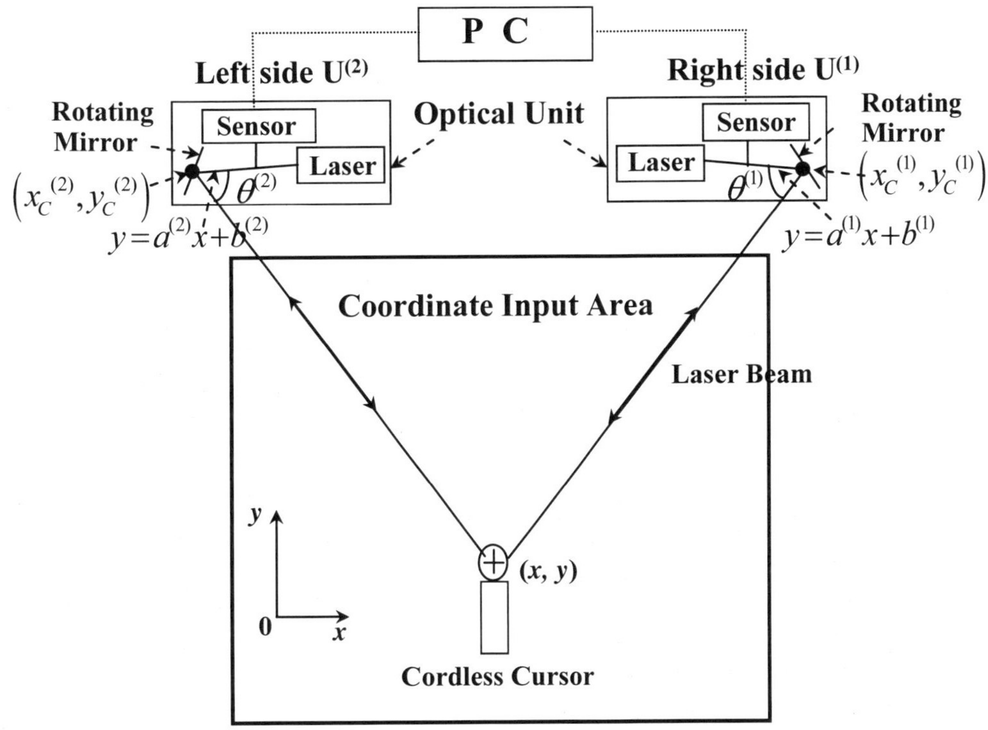

Figure 1. System Composition.

The cordless cursor can move in the coordinate input area of the digitizer. As both rotating plane mirrors rotate at about 2400 rpm, reflecting ray from this mirror scans the coordinate area.

Each main characteristics is as follows.

[1] Laser: Semiconductor laser, continuous wave, wavelength 780 nm, maximum outputs 30 mW.
[2] Motor: Brushless DC motor, 2400 rpm, rotating jitter (0.003 % or less).
[3] Sensor: Si-avalanche photo diode.
[4] Cursor: Bead type retro-reflective mirror.

If the scanning ray strikes the cursor,it is reflected, then, returns to the rotating mirror, and again is reflected,returns to the beam splitter and is reflected into the sensor all in negligible time. Then, a personal computer measures the time $t^{(j)}$ $(j=1,2)$ of ray sensing by the sensor, and the time $T^{(j)}$ $(j=1,2)$ required for mirror's one rotation.The angle $\theta^{(j)}$ of the rotating mirrors of $\boldsymbol{U}^{(j)}$ $(j=1,2)$ is calculated as the following equation.

$$\theta^{(j)} = 2\times 360 \times \frac{t^{(j)}}{T^{(j)}} \ (j=1,2). \tag{1}$$

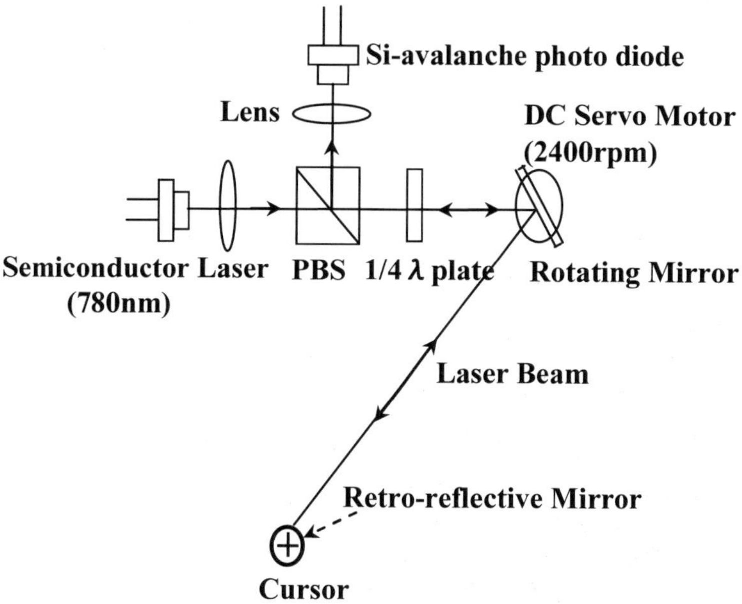

Figure 2. Optical Unit.

3.2.2. Problems of a Typical Coordinate Calculation Algorithm on Calibration

The position of the cursor can be calculated from both $\theta^{(j)}$ by the principle of triangle measurement.The typical coordinate calculation algorithm is described by

$$x = \frac{L\tan\theta^{(2)}}{\left(\tan\theta^{(1)} + \tan\theta^{(2)}\right)}, \tag{2}$$

$$y = \frac{L\tan\theta^{(1)}\tan\theta^{(2)}}{\left(\tan\theta^{(1)} + \tan\theta^{(2)}\right)}. \tag{3}$$

Where L is a distance between the centers of rotating mirrors.

Such a typical algorithm has the following serious problems on calibration:

[1] The ray from the laser needs to enter exactly to the center of rotating mirror.
[2] The distance L between the centers of rotating mirrors needs to be measured precisely.
[3] Each ray from the lasers needs to enter the center of the rotating mirror.
[4] The rotating center of the mirror needs to stay on the surface of the mirror.

Therefore, it is difficult to realize on-site calibration with the above coordinate calculation algorithm.

3.2.3. Appropriate Coordinate Calculation Algorithm for On-Site Calibration

Consider an appropriate model for calculating coordinates to solve the above problems of our system on calibration [2, 3].

Assume that ($x_C^{(j)}$, $y_C^{(j)}$) denotes the rotating center position of rotating mirror $\boldsymbol{U}^{(j)}$ and the equation of the ray from the laser of $\boldsymbol{U}^{(j)}$ is

$$y = a^{(j)}x + b^{(j)} \tag{4}$$

The angle of the rotating mirror is set to zero when the mirror is perpendicular to the ray entering from the laser.

Assume that the angle of the rotating mirror of $\boldsymbol{U}^{(j)}$ is $\theta^{(j)}$ when the position of the cursor is (x, y) and that the ray from the laser of $\boldsymbol{U}^{(j)}$ enters the rotating mirror at ($x_C^{(j)}$, $y_C^{(j)}$), (j=1, 2). And also assume that the refelected point on the rotating mirror of laser ray is ($x_I^{(j)}$, $y_I^{(j)}$).

Then, the following three relations can be obtained.

$$\left(y_I^{(j)} - y_C^{(j)}\right) = \tan\left(\theta^{(j)}/2 + \alpha - \pi/2\right)\left(x_I^{(j)} - x_C^{(j)}\right), \tag{5}$$

$$\left(y_I^{(j)} - y_C^{(j)}\right) = a^{(j)}\left(x_I^{(j)} - x_C^{(j)}\right) + b^{(j)}, \tag{6}$$

$$\arctan\frac{\left(y_I^{(j)} - y_C^{(j)}\right)}{\left(x_I^{(j)} - x_C^{(j)}\right)} + \pi - \alpha = \theta^{(j)}, \tag{7}$$

where $\alpha = \tan a^{(j)}$.

Therefore, the following equation are obtained for $\boldsymbol{U}^{(j)}$ (j=1,2) from Eqs. (5) ~(7).

$$\left(\cos\theta^{(j)} - a^{(j)}\sin\theta^{(j)}\right)\left(y - y_C^{(j)}\right) - \left(a^{(j)}\cos\theta^{(j)} + \sin\theta^{(j)}\right)\left(x - x_C^{(j)}\right) = b^{(j)}, \tag{8}$$

where the sign of angle $\theta^{(j)}$ is *+1* when *j=1* and *-1* when *j=2*.

If we measure $\theta^{(j)}$ for four individual cursor positions (we call these points control points hereafter) (X_n, Y_n) $(n=1,\dots,4)$ which are known beforehand, the unknown parameters $a^{(j)}$, $b^{(j)}$, $x_C^{(j)}$, $y_C^{(j)}$ can be calculated using Eq. (8).

Once those unknown parameters are calculated, any cursor position (x, y) in the coordinate area can be calculated by measuring $\theta_n^{(j)}$ and by using Eq. (8).

Nonlinear equation (8) can be solved algebraically as follows.

Assume that $\theta_n^{(j)}$ is measured for four known cursor positions (X_n, Y_n) $(n=1,\dots,4)$. Then, Eq.(8) is transformed into the following linear equation (9).

$$Q_n^{(j)}p^{(j)} + P_n^{(j)}q^{(j)} + \left(Y_nQ_n^{(j)} + X_nP_n^{(j)}\right)a^{(j)} + b^{(j)} = -X_nP_n^{(j)} + Y_nQ_n^{(j)} \tag{9}$$

by variable transformation [9] such as

$$p^{(j)} = -a^{(j)}x_C^{(j)} + y_C^{(j)}, \tag{10}$$

$$q^{(j)} = -x_C^{(j)} - a^{(j)}y_C^{(j)}, \tag{11}$$

$$Q_n^{(j)} = \cos\left(\theta_n^{(j)}\right), \tag{12}$$

$$P_n^{(j)} = \sin\left(\theta_n^{(j)}\right), \tag{13}$$

where $j=1,2$, and $n=1,2,\dots,4$.

Eq. (9) can be easily solved about variables $p^{(j)}$, $q^{(j)}$, $a^{(j)}$, $b^{(j)}$. Therefore, the unknown parameters $a^{(j)}$, $b^{(j)}$, $x_C^{(j)}$, $y_C^{(j)}$ can be determined by inverse variable transformation.

Suppose that

$$\mathbf{A}^{(j)} = \begin{bmatrix} Q_1^{(j)} & P_1^{(j)} & Y_1Q_1^{(j)} + X_1P_1^{(j)} & 1 \\ Q_2^{(j)} & P_2^{(j)} & Y_2Q_2^{(j)} + X_2P_2^{(j)} & 1 \\ Q_3^{(j)} & P_3^{(j)} & Y_3Q_3^{(j)} + X_3P_3^{(j)} & 1 \\ Q_4^{(j)} & P_4^{(j)} & Y_4Q_4^{(j)} + X_4P_4^{(j)} & 1 \end{bmatrix}, \tag{14}$$

$$\mathbf{c}^{(j)} = \begin{bmatrix} -X_1P_1^{(j)} + Y_1Q_1^{(j)} \\ -X_2P_2^{(j)} + Y_2Q_2^{(j)} \\ -X_3P_3^{(j)} + Y_3Q_3^{(j)} \\ -X_4P_4^{(j)} + Y_4Q_4^{(j)} \end{bmatrix}, \tag{15}$$

and

$$\mathbf{x}^{(j)} = \left(p^{(j)}, q^{(j)}, a^{(j)}, b^{(j)}\right). \tag{16}$$

Then, Eq. (9) can be shown by

$$\mathbf{A}^{(j)}\mathbf{x}^{(j)} = \mathbf{c}^{(j)}, \tag{17}$$

and so, four parameters $p^{(j)}$, $q^{(j)}$, $a^{(j)}$, $b^{(j)}$ can be obtained by

$$\mathbf{x}^{(j)} = \left(\mathbf{A}^{(j)T}\mathbf{A}^{(j)}\right)^{-1}\mathbf{A}^{(j)T}\mathbf{c}^{(j)}, \tag{18}$$

when the coefficient matrix $\mathbf{A}^{(j)}$ is regular, i.e.

$$\det\left(\mathbf{A}^{(1)}\mathbf{A}^{(2)}\right) \neq 0. \tag{19}$$

Therefore, four parameters $a^{(j)}$, $b^{(j)}$, $x_C^{(j)}$, $y_C^{(j)}$ can be determined by Eqs. (10), (11) and (18).

Finally,we can obtain the following coordinate calculation equations.

$$\begin{pmatrix} y \\ x \end{pmatrix} = \begin{pmatrix} a_1 & b_1 \\ a_2 & b_2 \end{pmatrix}^{-1} \begin{pmatrix} c_1 + a_1 y_C{}^{(1)} + b_1 x_C{}^{(1)} \\ c_2 + a_2 y_C{}^{(2)} + b_2 x_C{}^{(2)} \end{pmatrix}, \tag{20}$$

where

$$a_j = \cos\theta^{(j)} - a^{(j)}\sin\theta^{(j)}, \tag{21}$$

$$b_j = -\sin\theta^{(j)} - a^{(j)}\cos\theta^{(j)}, \tag{22}$$

$$c_j = b^{(j)}. \tag{23}$$

3.3. Error Analysis

In this section, the error analysis about angle which makes coordinate error smaller is mentioned.

The coordinate error differs depending on the four control points (X_n, Y_n) $(n=1,\ldots,4)$ because of the error of the angle measured by cursor. Therefore, the following simulation was done.

[1] Assume that the coordinates area is 3900 mm × 2700 mm and the origin of xy coordinate stays at the left bottom point, x axis is horizontal and y axis is vertical.
[2] Assume that the coordinate values of four control points are known exactly.
[3] Assume that the average value and standard deviation of any angle error are zero and 0.0005 degree, respectively.
[4] The evaluation of coordinate error for various different of four control points set is made as follows. Set error to 12 angles by random numbers and calculate coordinate error $dx_{mn}{}^{k}$ and $dy_{mn}{}^{k}$ at the center points of 13 × 9 equally divided small grids i.e. small area of 30 mm × 30 mm.

And this process recurs K times and calculates the following variance and covariance.

$$\delta_{dx_{mn}}^2 = \frac{1}{K}\sum_{k=1}^{K}\left(dx_{mn}^k\right)^2, \tag{24}$$

$$\delta_{dy_{mn}}^2 = \frac{1}{K}\sum_{k=1}^{K}\left(dy_{mn}^k\right)^2, \tag{25}$$

$$\delta_{dx_{mn}dy_{mn}} = \frac{1}{K}\sum_{k=1}^{K}\left(dx_{mn}^k dy_{mn}^k\right), \tag{26}$$

where $K = 120$.

The error at center point of each grid is shown by error ellipse.
The error is shown by

$$(x, y)\sum_{mn}^{-1}\begin{pmatrix} x \\ y \end{pmatrix} = c \ \left(c \cdots \text{constant}\right), \tag{27}$$

where Σ is variance-covariance matrix described by

$$\sum_{mn} = \begin{pmatrix} \delta_{dx_{mn}}^2 & \delta_{dx_{mn}dy_{mn}} \\ \delta_{dx_{mn}dy_{mn}} & \delta_{dy_{mn}}^2 \end{pmatrix}. \tag{28}$$

Fig. 3 shows the coordinate error diagrams in case of four control points being on the four corners of the coordinate area, which is obtained by the simulation for error analysis.

Since the coordinate error is largely different according to four control points, we make an optimal simulation to determine the position of four control points which makes the summation of the variances $\delta_{dx_{mn}}^2 + \delta_{dy_{mn}}^2$ calculated at the center points of grid points described by Eqs. (24) and (25) minimize in section 3.4.

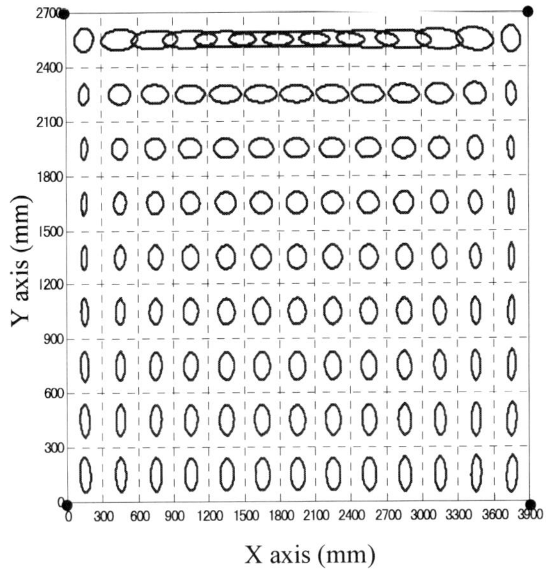

Four control points are at the four corners of the area defined by ●.

Figure 3. Coordinate error diagram of 4 m × 3 m digitizer obtained by simulation described by ellipse.

3.4. Optimization of Four Control Points

3.4.1. Appropriate Model for Determining the Optimal Four Control Points [10]

Let consider the deviations of parameters in fundamental equation Eq. (8)

$\Delta x_C{}^{(j)}$, $\Delta y_C{}^{(j)}, \Delta b^{(j)}$, $\Delta a^{(j)}$, $\Delta\theta_k^{(j)}, \Delta x^{(j)}, \Delta y^{(j)}$ $(j=1,2; k=1,2,...,5)$. Then, the following relations in these parameters can be determined by carrying out total differential with respect to the Eq. (8)

$$\begin{aligned}
&\left(a^{(j)}\cos\theta_k^{(j)}+\sin\theta_k^{(j)}\right)\Delta x_C{}^{(j)}-\left(\cos\theta_k^{(j)}-a^{(j)}\sin\theta_k^{(j)}\right)\Delta y_C{}^{(j)}\\
&\Delta b^{(j)}-\left(\sin\theta_k^{(j)}\left(y_k-y_C{}^{(j)}\right)+\cos\theta_k^{(j)}\left(x_k-x_C{}^{(j)}\right)\right)\Delta a^{(j)}\\
&+\left(\left(a^{(j)}\sin\theta_k^{(j)}-\cos\theta_k^{(j)}\right)\left(x_k-x_C{}^{(j)}\right)-\left(\sin\theta_k^{(j)}+a^{(j)}\sin\theta_k^{(j)}\right)\left(y_k-y_C{}^{(j)}\right)\right)\Delta\theta_k^{(j)}\\
&-\left(a^{(j)}\cos\theta_k^{(j)}+\sin\theta_k^{(j)}\right)\Delta x^{(j)}+\left(\cos\theta_k^{(j)}-a^{(j)}\sin\theta_k^{(j)}\right)\Delta y^{(j)}\\
&=0\ \ (j=1,2; k=1,2,...,5)
\end{aligned} \tag{29}$$

Each coefficient of $\Delta x_C{}^{(j)}, \Delta y_C{}^{(j)}, \Delta b^{(j)}, \Delta a^{(j)}, \Delta\theta_k^{(j)}, \Delta x^{(j)}, \Delta y^{(j)}$ in Eq. (29) is denoted by

$Cx_C{}^{(j)}(k), Cy_C{}^{(j)}(k), -1, Ca^{(j)}(k), C\theta^{(j)}(k), Cx^{(j)}(k), Cy^{(j)}(k)$, respectively, and so, Eq. (29) can be rewritten by

$$\begin{aligned} & Cx_C{}^{(j)}(k)\Delta x_C{}^{(j)} - Cy_C{}^{(j)}(k)\Delta y_C{}^{(j)} - \Delta b^{(j)} + Ca^{(j)}(k)\Delta a^{(j)} \\ & = -C\theta^{(j)}(k)\Delta\theta_k^{(j)} - Cx^{(j)}(k)\Delta x^{(j)} - Cy^{(j)}(k)\Delta y^{(j)} \end{aligned} \tag{30}$$

Then, Eq. (30) can be expressed by a matrix form as

$$\mathbf{AX} = \mathbf{C}, \tag{31}$$

where

$$\mathbf{A} = \begin{pmatrix} \mathbf{sA}^{(1)} & \mathbf{0} \\ \mathbf{0} & \mathbf{sA}^{(2)} \end{pmatrix}, \tag{32}$$

$$\mathbf{sA}^{(j)} = \begin{pmatrix} Cx_C{}^{(j)}(1) & Cy_C{}^{(j)}(1) & Ca^{(j)}(1) & -1 \\ Cx_C{}^{(j)}(2) & Cy_C{}^{(j)}(2) & Ca^{(j)}(2) & -1 \\ Cx_C{}^{(j)}(3) & Cy_C{}^{(j)}(3) & Ca^{(j)}(3) & -1 \\ Cx_C{}^{(j)}(4) & Cy_C{}^{(j)}(4) & Ca^{(j)}(4) & -1 \end{pmatrix}, \tag{33}$$

$$\begin{aligned} \mathbf{X} &= \left(x_1,\ x_2, x_3, x_4, x_5, x_6, x_7, x_8\right)^T \\ &= \left(\Delta x_C{}^{(1)}, \Delta y_C{}^{(1)}, \Delta a^{(1)}, \Delta b^{(1)}, \Delta x_C{}^{(2)}, \Delta y_C{}^{(2)}, \Delta a^{(2)}, \Delta b^{(2)}\right)^T, \end{aligned} \tag{34}$$

$$\mathbf{C} = -\begin{pmatrix} C\theta^{(1)}(1)\Delta\theta_1^{(1)} + Cx^{(1)}(1)\Delta x_1^{(1)} + Cy^{(1)}(1)\Delta y_1^{(1)} \\ C\theta^{(1)}(2)\Delta\theta_2^{(1)} + Cx^{(1)}(2)\Delta x_2^{(1)} + Cy^{(1)}(2)\Delta y_2^{(1)} \\ C\theta^{(1)}(3)\Delta\theta_3^{(1)} + Cx^{(1)}(3)\Delta x_3^{(1)} + Cy^{(1)}(3)\Delta y_3^{(1)} \\ C\theta^{(1)}(4)\Delta\theta_4^{(1)} + Cx^{(1)}(4)\Delta x_4^{(1)} + Cy^{(1)}(4)\Delta y_4^{(1)} \\ C\theta^{(2)}(1)\Delta\theta_1^{(2)} + Cx^{(2)}(1)\Delta x_1^{(2)} + Cy^{(2)}(1)\Delta y_1^{(2)} \\ CO^{(2)}(2)\Delta\theta_2^{(2)} + Cx^{(?)}(2)\Delta x_2^{(?)} + Cy^{(2)}(2)\Delta y_2^{(2)} \\ C\theta^{(2)}(3)\Delta\theta_3^{(2)} + Cx^{(2)}(3)\Delta x_3^{(2)} + Cy^{(2)}(3)\Delta y_3^{(2)} \\ C\theta^{(2)}(4)\Delta\theta_4^{(2)} + Cx^{(2)}(4)\Delta x_4^{(2)} + Cy^{(2)}(4)\Delta y_4^{(2)} \end{pmatrix}. \tag{35}$$

Therefore, the deviations of parameters $\Delta x_C{}^{(j)}, \Delta y_C{}^{(j)}, \Delta b^{(j)}, \Delta a^{(j)}$ can be determined by

$$\mathbf{X} = \left(\mathbf{A}^{\mathbf{T}}\mathbf{A}\right)^{\mathbf{-1}} \mathbf{A}^{\mathbf{T}}\mathbf{C} \tag{36}$$

from Eq. (34).

Then, the deviations $\Delta x, \Delta y$ of coordinates x, y can be determined by

$$\begin{pmatrix} \Delta y \\ \Delta x \end{pmatrix} = \begin{pmatrix} \Delta a_1 & \Delta b_1 \\ \Delta a_2 & \Delta b_2 \end{pmatrix}^{-1} \begin{pmatrix} \Delta c_1 + \Delta a_1 \Delta y_C{}^{(1)} + \Delta b_1 \Delta x_C{}^{(1)} \\ \Delta c_2 + \Delta a_2 \Delta y_C{}^{(2)} + \Delta b_2 \Delta x_C{}^{(2)} \end{pmatrix}, \tag{37}$$

where

$$\Delta a_j = \cos\Delta\theta^{(j)} - \Delta a^{(j)} \sin\Delta\theta^{(j)}, \tag{38}$$

$$\Delta b_j = -\sin\Delta\theta^{(j)} - \Delta a^{(j)} \cos\Delta\theta^{(j)}, \tag{39}$$

$$\Delta c_j = \Delta b^{(j)}. \tag{40}$$

From Eqs. (31) ~ (40), these are also expressed by the linear combinations of angle deviations $\Delta\theta_k^{(j)}$ $(j = 1,2; k = 1,2,...,5)$ of both right and left side angles measured by sensors as

$$\Delta x = \sum_{j=1}^{2} \sum_{k=1}^{5} D(\theta)_k^{(j)} \Delta\theta_k^{(j)}, \tag{41}$$

$$\Delta y = \sum_{j=1}^{2} \sum_{k=1}^{5} E(\theta)_k^{(j)} \Delta\theta_k^{(j)}, \tag{42}$$

Then, the variances $\delta_{\Delta x}{}^2$, $\delta_{\Delta y}{}^2$ of coordinate deviations Δx, Δy can be described by

$$\delta_{\Delta x}{}^2 = \sum_{j=1}^{2} \sum_{k=1}^{5} \left[D(\theta)_k^{(j)} \Delta\theta_k^{(j)} \right]^2, \tag{43}$$

$$\delta_{\Delta y}{}^2 = \sum_{j=1}^{2} \sum_{k=1}^{5} \left[E(\theta)_k^{(j)} \Delta\theta_k^{(j)} \right]^2, \tag{44}$$

because the angle deviations $\Delta\theta_k^{(j)}$ $(j = 1,2; k = 1,2,...,5)$ are dependent each other.

Then, the model for determining the optimal four control points is assumed to be described by

$$\min_{(\mathbf{x},\mathbf{y}) \in S} F(\mathbf{x}, \mathbf{y}), \tag{45}$$

where

$$F(\mathbf{x},\mathbf{y}) = \int_S \left(\delta_{\Delta x}{}^2 + \delta_{\Delta y}{}^2\right)dxdy$$
$$= \sum_{j=1}^{2}\sum_{k=1}^{5}\int_S \left(D(\theta)_k^{(j)2} + E(\theta)_k^{(j)2}\right)\Delta\theta_k^{(j)2}dxdy \quad (46)$$

Here, $(\mathbf{x},\mathbf{y}) = ((x_1, x_2, x_3, x_4),(y_1, y_2, y_3, y_4))$ denotes four control points, and S denotes the coordinate area of this digitizer system and it is assumed to be

$$S = ((x, y) : 0 \le x \le l_x, 0 \le y \le l_y). \quad (47)$$

Here, the integral value in Eq. (46) is determined approximately by the following numerical calculation.

$$F(\mathbf{x},\mathbf{y}) = \sum_{j=1}^{2}\sum_{k=1}^{5}\sum_{m=1}^{n}\left(D(\theta)_k^{(j)2} + E(\theta)_k^{(j)2}\right)\Delta\theta_k^{(j)2}S_m, \quad (48)$$

where

$$S = \sum_{m=1}^{n} S_m,\ S_i \cap S_j = \phi(i \ne j). \quad (49)$$

Namely, the problem for determining the optimal four control points from the accuracy point of view is reduced to the following constrained optimization problem.

Objective function:

$$F(\mathbf{x},\mathbf{y}) = \sum_{j=1}^{2}\sum_{k=1}^{5}\sum_{m=1}^{n}\left(D(\theta)_k^{(j)2} + E(\theta)_k^{(j)2}\right)\Delta\theta_k^{(j)2}S_m. \quad (50)$$

Constraint condition:

$$x_1 = x_4,\ x_2 = x_3,\ y_1 = y_4,\ y_2 = y_3, \quad (51)$$

$$0 \le x_1 \le x_2 \le l_x,\ 0 \le y_1 \le y_3 \le l_y. \quad (52)$$

3.4.2. Method for Optimization and its Result

Since the objective function Eq. (50) is not convex function, it has more than one local minimal point. Therefore, the result of the optimization algorithm depends on the initial point and does not generally converge to the minimum point. To avoid the convergence to local minimal points, we carry out the following variable transformation

$$X_1 = \sqrt{x_1},\ X_2 = \sqrt{y_1}, X_3 = \sqrt{l_x - x_2}, X_4 = \sqrt{l_y - y_3} \tag{53}$$

and the optimal point is determined using the extended Lagrange multiplier method, which is one of the conventional methods of optimization [11].

Here, the extended Lagrange function is described by

$$L(X_1,\ X_2, X_3, X_4, \mathbf{u}, \mathbf{r}) = F(\mathbf{x}, \mathbf{y}) + \sum_{k=1}^{6} \frac{1}{2r_k}\left[\left(\max\left(0, u_k + r_k + g_k\right) - u_k\right)^2 - u_k{}^2\right] \tag{54}$$

where

$\mathbf{u} = \left(u_1, \cdots, u_6\right)$, and $\mathbf{r} = \left(r_1, \cdots, r_6\right)$ denotes Lagrange multiplier and penalty parameter, respectively. The inequality functions: $g_k \leq 0\left(k = 1, \cdots, 6\right)$ are inequality constraint functions.

The constraint optimization of the extended Lagrange function $L\left(X_1,\ X_2, X_3, X_4, \mathbf{u}, \mathbf{r}\right)$ is carried out by well-known quasi-Newton method [12].

In this chapter, the following assumptions on parameters are done considering laser digitizer system.

[1] Coordinate area S : the rectangular area of 3900 mm ×2700 mm.
[2] Each small area S_m (m=1,…, n) of S : the square area of 30 mm ×30 mm.
[3] Four parameters $a^{(j)}$, $b^{(j)}$, $x_C{}^{(j)}$, $y_C{}^{(j)}$: -0.2, -0.8, 3900, 2800 (j=1), and 0.2, -0.8, 0, 2800 (j=2).
[4] The distribution of measuremed angles follows the gaussian one and its standard deviation of the measurement angle: 0.0005 deg.
[5] Initial points of four control points (unit: mm):

a) (0, 0), (3900, 0), (0, 2700), (3900, 2700)
b) (975, 675), (2925, 675), (975, 2025), (2925, 2025)
c) (1300, 900), (2600, 900), (1300, 1800), (2600, 1800)
d) (0, 0), (1950, 0), (0, 1350), (1950, 1350)
e) (1950, 0), (3900, 0), (1950, 1350), (3900, 1350)
f) (0, 1350), (1950, 1350), (0, 2700), (1950, 2700)
g) (1950, 1350), (3900, 1350), (1950, 2700), (3900, 2700)

Based on the above assumptions, the optimazation calculation was carried out. As a result, any initial four control points, which are described by (a) ~ (g) of (5) in the above assumptions, converge to the same four control points that are expressed as (720.6, 515.0), (3179.4, 515.0), (3179.4, 2700), (720, 2700), and in this case, the average coordinate error was (dx, dy) = (0.121, 0.109) (mm). In fact, the average coordinate error was (dx, dy) = (0.262, 0.154) (mm) when the control points are (0, 0), (3900, 0), (0, 2700), (3900, 2700).

Here, the average coordinate error (*dx, dy*) is assumed to be determined by

$$\left(dx,dy\right)=\frac{1}{120\times13\times9}\left(\sum_{1\le k\le120}\ \sum_{1\le m\le13,1\le n\le9}\left|dx_{mn}^{\ k}\right|,\ \sum_{1\le k\le120}\ \sum_{1\le m\le13,1\le n\le9}\left|dy_{mn}^{\ k}\right|\right), \tag{55}$$

on the assumption (3) in section 3.3.

Fig. 4 shows the coordinate error diagrams in case of four control points being at the optimal points.

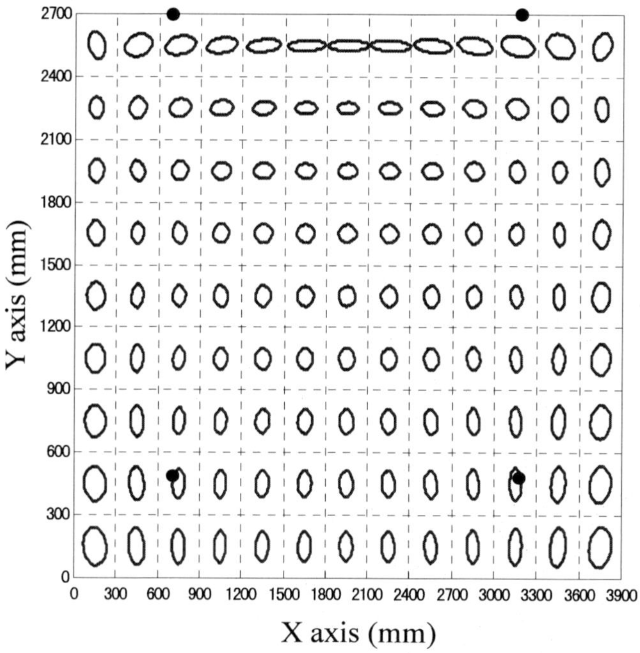

Four control points are at the optimal ones defined by ●.

Figure 4. Coordinate error diagram of 4 m×3 m digitizer obtained by simulation.

3.5. Experiment And Result

In order to estimate the accuracy of our system we made an experiment with the following procedure;

[1] Measure the both angles by using a cordless cursor for two dimensional grid points, positions of which are known exactly, all over the coordinate area of the digitizer.

[2] Determine the unknown four parameters of Ep. (18) using the grid points at the four corners and the optimal four control points of the area detrmined in section.4.

[3] Calculate the coordinates at *n* grid points according to our method, and this process recurs *m* times. Here, the caluculated coordinates at the grid points are assumed to be

denoted by $\left(x_k^{\ j}, y_k^{\ j}\right)$ $\left(k=1,..,n,\ j=1,2,...,m\right)$, and $(X_k^{j\prime}, Y_k^{j\prime})$ is corresponding exact coordinate,which is determined by a laser lemgth measuring machine.

[4] Calculate each deviation $(\Delta x_k^j, \Delta y_k^j)$ at each grid point $k=1,2,...,n$

$$\left(\Delta x_k^{\ j}, \Delta y_k^{\ j}\right)=\left(x_k^{\ j}-X_k^{\ j},\ y_k^{\ j}-Y_k^{\ j}\right). \tag{56}$$

[5] Calculate the variance $\delta^2_{\Delta xk}$, $\delta^2_{\Delta yk}$ and covariance $\delta_{\Delta xk\Delta yk}$ by

$$\delta^2_{dx_k}=\frac{1}{m-1}\sum_{j=1}^{m}\left(\Delta x_k^j\right)^2, \tag{57}$$

$$\delta^2_{dy_k}=\frac{1}{m-1}\sum_{j=1}^{m}\left(\Delta y_k^j\right)^2, \tag{58}$$

$$\delta_{dx_k dy_k}=\frac{1}{m-1}\sum_{j=1}^{m}\left(\Delta x_k^j \Delta y_k^j\right). \tag{59}$$

[6] Describe the coordinate measurement error at each grid point by

$$(x,y)\sum_k{}^{-1}\begin{pmatrix}x\\y\end{pmatrix}=c\ \left(c\cdots\text{constant}\right), \tag{60}$$

where Σ is variance-covariance matrix described by

$$\sum_k=\begin{pmatrix}\delta^2_{dx_k} & \delta_{dx_k dy_k}\\ \delta_{dx_k dy_k} & \delta^2_{dy_k}\end{pmatrix}. \tag{61}$$

[7] Average coordinate measurement errors described in this scetion are assumed to be calculated from Eq. (56) by

$$\left(dx,dy\right)=\frac{1}{nm}\left(\sum_{1\le k\le n}\sum_{1\le j\le m}\left|\Delta x_k^{\ j}\right|, \sum_{1\le k\le n}\sum_{1\le j\le m}\left|\Delta y_k^{\ j}\right|\right) \tag{62}$$

on the assumption that the measurement uncertainties of the both coordinates of four control points and grid points $\left(x_k^{\ j}, y_k^{\ j}\right)$ $\left(k=1,..,n,\ j=1,2,...,m\right)$ measured by a laser length measuring machine are all negligible. Variables n and m are assumed to be 117(=13*9) and 7, respectively.

Fig. 5(a) and (b) show the coordinate mesurement error diagrams by experiment in case of four control points being at the four corners of the coordinates areas and at optimal position, respectively.

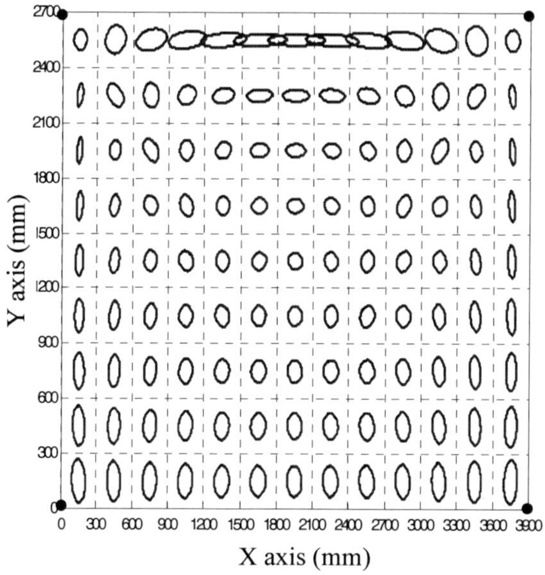

(a) Four control points are at the four corners of the area defined by ●.

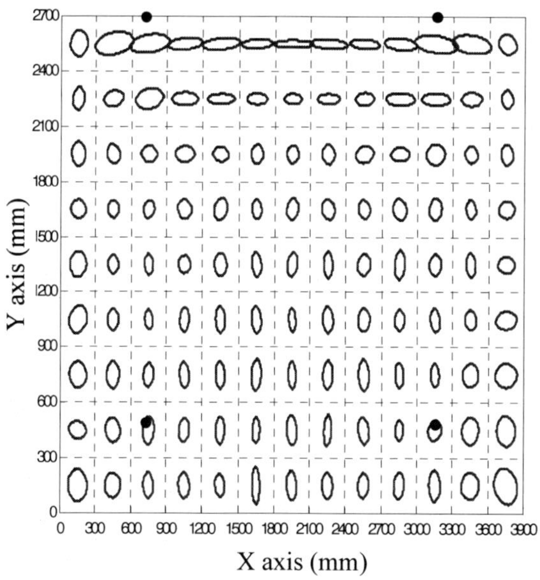

(b) Four control points are at the optimal ones defined by ●

Figure 5. Coordinate measurement error diagram of 4 m×3 m digitizer obtained by experiment described by ellipse.

Though the average coordinate error was about 0.14 mm in case of four control points being at the corners of the coordinate area, it was about 0.12 mm in case of four control points being at the optimal ones.

Therefore, the better four control points can be obtained by error analysis simulation. In case that the size of x-axis is largely different from that of y-axis, the error in case of four control points being at the optimal positions is largely improved. For example, in case of 6 m ×2 m digitizer which is used for car design, the average errors in case of four control points being on the straght lines, at the four corners, and at the optimal positions are 0.70, 0.22, and 0.14 mm, respectively [6].

In this way, we can realize the on-site calibration of a large sclae laser digitizer system with on-site insurance of measurement accuracy.

3.6. Conclusion

In this chapter, we proposed the on-site calibration method of a large scale laser digitizer in the measuring filed with error analysis and basic experiment. This method can be made possisble by the following properties.

[1] Architecture: the triangulation system using laser-ray which can make concentrate the equipments requiring geometrical accuracy into small number of small units.
[2] Algorithm: linealized algorithm which is invariant with alignment mismatch and mutual distance inaccuracy.

Especially, we proposed the linealized algoritm and the simulation for error analysis with connection to the angles measured at the four control points that are neccesary to determine coordinate calculation algorithm, and the optimal simulation for determining the position of four control points to give higher accuracy in rather detail.

In this way, we realized the on-site calibration and on-site insurance of measurement accuracy. The time required for on-site calibration is within 5 minutes and the average coordinate measurement error of 4 m×3 m digitizer is about 0.12 mm.

In this chapter, the evaluation of coordinate calculated values was carried out only by conventional error analysis and it is not completed because the position error of control point is not considered and the evaluation of measurement uncertainty of angle is not enough,etc. Therefore, we will proprose the uncertainty analysis of the laser digitizer system in detail based on the varinance-covariance matrix detrmined by Eq. (41) ~ (44) in the near future.

References

[1] Hearn E. D, Lankford L. G. *Digitizer Source Book*. Geocomp Ltd., 1989.
[2] Murakami A, Yamamoto S, Yamanami T. Development of Coordinate Measuring Equipment using electromagnetic give and receive action method. In *Proceedings of CPEM*.1988:283-4.

[3] Murakami A, Hosaka K, Fukushima M, Maeda M, Tuya N. An amorphous magnetostrictive delay line cordless digitizer. *IEEE Trans. Magn.* 1989;25;3:2739-44.
[4] Yatagai T. *Applied Optics: Introduction of Optical Measurement.* Maruzen Publishers, 1988 (in Japanese).
[5] Saito N, Hirose T, Abe M, Suwama M, Fujimoto I, Koizumi S,Yamene R, Murakami A. Development of super large size laser digitizer. In *Proceedings of Korea Automatic Contorol Conference*. 1990;2:1501-9.
[6] Fujimoto I, Hirose T, Abe M. A two dimensional large scale laser digitizer. In *Proceedings of IEEE instrumentation and measurement technology conference (IMTC 93).* 1993:416-9.
[7] Fujimoto I, Hirose T, Abe M, Saito N,Takahashi D. *Two dimensional Large Scale Laser Digitizer System with a Cordless Cursor.* The Institute of Systems, Control and Information Engineers of Japan. 1993;6;11:479-487 (in Japanese).
[8] Fujimoto I, Abe M. *Application of Two Dimensional Laser Digitizer to Surveying.* The Institute of Systems, Control and Information Engineers of Japan. 1994;7;9:391-3 (In Japanese).
[9] Mathematical Society of Japan, *Mathematics Dictionary (4th Edition Version): III-5 Algebraic Equation.* Iwanami Shoten Publishers, 2007.
[10] Fujimoto I. The Method of Initial Experiment in Laser Digitizer. The Institute of Electronics, *Information and Communication Engineer of Japan* D-II.1995; J78;12:1859-67 (in Japanese).
[11] Fukushima M. *Nonlinear Optimization Theory.* Sangyo Toshyo Publishers. 1981 (in Japanese).
[12] Ibaraki T, Fukushima M. *Optimization Programing.* Iwanami shoten publishers. 1985:167-207 (in Japanese).

In: Calibration Technology, Theories and Applications ISBN: 978-1-62618-808-2
Editors: Ikumatsu Fujimoto and Kunitoshi Nishimura

Chapter 4

Calibration for Radiation Assessment in the Multi-group Approximation

Ateba P. Owono*

Laboratoire de Physique Nucléaire, Université de Yaoundé I, Cameroon

Abstract

This chapter introduces the calibration of a nuclear radiation detection system. It starts with a recapitulation of the history of radioactivity, its discovery, nature and properties, the nuclear electronics and the methods deployed for its analysis; it then presents a particular analysis method for gamma rays in which the detection system using a NaI(Tl) scintillation detector, is calibrated with m different radionuclides. The observed spectrum from any sample is hereupon subdivided into exactly m regions considered as energy groups, which enables the use of Multi-group techniques leading to a square matrix of m^2 elements for the calculation of the activities of the radionuclides of the sample. The chapter further examines the big problem of calculating the uncertainty, since the method uses matrices, and delivers first conclusions: The uncertainty of the assessed activity is found to be specifically tied to that of the calibrating activity and, in terms of matrices, the relative uncertainty of the inverse matrix appears to be the transposed matrix to the relative uncertainty of the original matrix. This should be a mathematical theorem. In fact, both result schemes easily apply to solving any linear problem by means of matrices.

Keywords: Environmental assessment, gamma spectroscopy, photon scattering coefficients, sensitiveness, calibration coefficients, scintillation detector, nuclear electronics, standards, regression line, error propagation, uncertainty estimation.

4.1. Introduction

Matter is made of atoms. Some atomic nuclei spontaneously decay on their own by emitting rays called radiation, a phenomenon that is named "radioactivity". In a first stage, radioactivity analysis aims at delivering the activity corresponding to a given number of counts registered by a radiation counter. The counting device must therefore be calibrated with a radioactive source of known activity to get the corresponding number of counts. Counting a sample of unknown activity with the calibrated device then yields the corresponding activity of the sample.

Each ray has a given energy. A picture depicting the count number as a function of the rays' energy, is called the energy spectrum of the sample. A ray emitting nucleus or radionuclide generally emits rays of definite energies. The energy spectrum of a radionuclide therefore has, according to those definite energies, one or more spectral lines or peaks that characterise that radionuclide and enable its identification. So in a second stage usually carried out in a laboratory, radioactivity analysis identifies the decaying radionuclides of the sample through their characteristic spectral lines and assesses their quantity from registered counts.

One part of the rays generating a spectral line of a radionuclide is totally absorbed in the detector and counted at the level of the corresponding energy on the spectrum, giving the so called "total absorption peak" (TAP). The other part eventually associated to other phenomena, is scattered over the whole of energy spectrum and partially counted there. Traditional methods analyse just the TAP. So the other rays counted elsewhere on the spectrum are not taken into account. This is a loss and therefore a weakness of such methods. To overcome this weakness, a new way is being studied for a couple of years now. This chapter is a contribution to that study.

Discovered at the end of the 19th century, radioactivity and, in general nuclear radiation, has been largely investigated and recognized as a very useful tool for mankind. It is widely deployed nowadays in Medicine and Industry and, indeed, in nearly all scientific disciplines. In many cases, it delivers information available by no other means [Fisher et al.]. But it nevertheless bears some danger for living organisms and must therefore be handled with care. Ambient radiation exists everywhere on Earth. To know and master it well, and also for protection purpose, man must measure and assess it with good precision, qualitatively and quantitatively. This legitimates further research on improving radiation detection systems and methods, which make new device calibration necessary.

According to the *Vocabulaire International de Métrologie* (VIM) which is internationally recognized as the document providing technical definitions on calibration and other measurement related terms [Payne] (sic):

- "Calibration is the process of determining the relationship between the readings obtained by a measuring instrument or system and the applicable units of some defined system of measurement;
- 'Calibration quantifies the relationship between the readings of a measuring instrument (or system) and the relevant units in the SI system;
- 'Calibration is essentially a process of comparison. An instrument is used to measure (...) a calibration standard, and the result is compared to two things: the known value and uncertainty of the standard and the performance specifications required by the customer".

Basing on that concept, this chapter presents the calibration of a radiation measurement system relatively to two measurands: the ray energy and, as a new contribution, the radionuclide activity in the multi-group approximation, both with a particular accent laid on the expected uncertainty assessment.

4.2. Radioactivity and the Atomic Nucleus

The discovery of radioactivity is a passionate story, as can be felt while reading different authors [Mann et al.]. Just the essential of it be reported below, to let the reader have a glance at how nuclear radiation came to attention of man.

In 1895, the German physicist Wilhelm Conrad Roentgen (1845 – 1923) discovered that a glass surface that is hit by cathode rays fluoresces and emits particularly penetrating rays that he named X-rays. Roentgen could take a photograph of either the skeleton of the hand of a living person, or the content of a hermetically closed wooden box. One year later, in 1896, French physicist Antoine Henri Becquerel (1852-1908) wanted to find out if each phosphorescent substance emits X rays. This led him to discover that also Uranium sulphate, although not exposed to sunlight or to any other radiation, permanently emits highly penetrating rays. Becquerel hereupon discovered the natural radioactivity of Uranium. His works were followed by those of Marie Sklodowska-Curie (1867–1934) and her husband Pierre Curie (1859-1906), who discovered the natural radioactivity of Thorium, Radium and Polonium.

In 1899, Ernest Rutherford (1871-1937) established that there are three types of rays occurring in natural radioactivity: alpha rays of positive electric charge, beta rays of negative charge and gamma rays without electric charge. He and his compatriot Frederick Soddy (1877 –1956) proposed as explanation for natural radioactivity, that some atoms are unstable and can spontaneously decay by emitting energy in the form of radiation, hereupon transforming themselves into atoms of other chemical elements. So matter contains energy it can emit. In 1905, German physicist Albert Einstein (1879-1955) established the mass-energy equivalence formula.

The evaluations of a lot of radioactivity experiments on a large variety of substances finally made Rutherford in 1911 propose an atomic model consisting in a dense centre and a periphery, the large space between both being empty to allow particles to go through. Very far away from the centre, the periphery is beset with very fast moving electrons. Upon naming "nucleus" the centre which he assumed to nearly carry the whole of atomic mass, Rutherford founded Nuclear Physics in 1911! According to him, the nucleus is composed of tiny particles of almost 2000 times the mass of an electron and a positive elementary electric charge each. He named them "protons" and stated their number to be equal to that of the electrons of the periphery. It was given the name "atomic number" and defines the chemical element. In 1932, James Chadwick (1891 - 1974) discovered that the nucleus also shelters other small particles of almost the same mass like the proton each, but of no electric charge, the neutrons. Together, protons and neutrons are called "nucleons", as constitutive particles of the nucleus. The number of nucleons of an atomic nucleus is called mass number. Atoms with same atomic number but different mass numbers are called isotopes, and radioisotopes if they undergo radioactive decay.

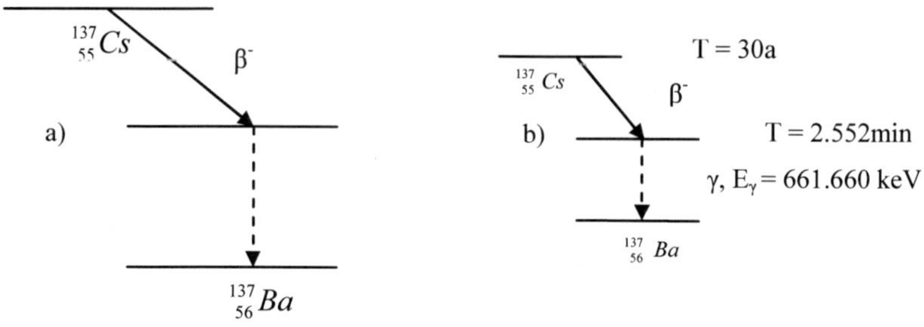

Figure 1. Decay scheme of Cesium isotope 137.

Figure 1 depicts a decay scheme. Through radioactive β^- decay (Figure 1a), Cesium 137 that has 55 protons and a mass number of 137 nucleons turns into Barium 137 that has 56 protons. The atomic number increases here by one unit, moving towards the chemical element right from Cesium; therefore the direction of the bold arrow. So, nuclear decay is the first property of radioactivity. As second property, the population of identical radionuclides at a given date decreases with ongoing time. The time T after which it has lost the half of its members is called the "half-life" of the radionuclide. Figure 1b shows Cesium137 that has a half-life of thirty years. Barium 137 is produced in an unstable or excited state that has a half-life of 2.552 minutes and an excitation energy of 661.660keV that de-excites (discontinued vertical arrow) into the fundamental state by emitting a gamma ray that has an energy of 661.660 keV [Browne et al.], see also Figure 5.

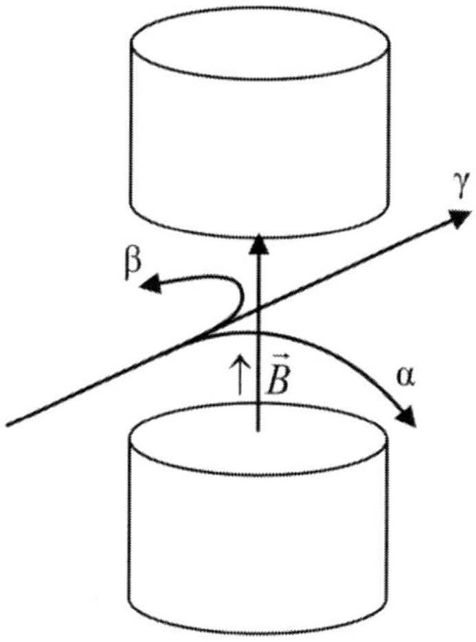

Figure 2. Rays emitted by radioactivity deflect right (α), left (β) or not at all (γ) across an electromagnetic field ($\vec{B}$).

The three kinds of invisible rays emitted by radioactive substances were simply named after the first three characters of the Greek alphabet: alpha rays, beta rays and gamma rays, depending on whether they are deflected to the right or to the left, or not at all, while crossing an electromagnetic field (Figure 2). This ray deflection may be considered as the third property of radioactivity. A particularity is that there are two kinds of beta decays: the beta plus and the beta minus decays of which the theoretical explanation is depicted on Figure 3 over the change of atomic numbers.

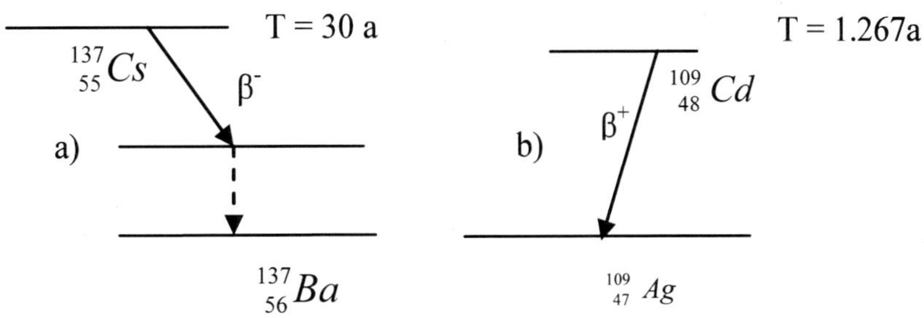

Figure 3. a) β^- decay (Cesium 137); b) β^+ decay: Cadmium 109 turns into Silver 109 directly in the fundamental or ground state by decreasing its atomic number by 1 unit from 48 to 47. Silver is the left neighbour to Cadmium, therefore the direction of the bold arrow.

Although invisible to man's eye, gamma-rays are known to be of same nature like visible light and X-rays: they are electromagnetic waves and are not deflected while crossing an electromagnetic field. They cause similar effects on matter: too much sun light destroys skin, so do gamma-rays, X-rays, and also particle rays. One therefore must protect oneself against radiation. As fourth property of radioactivity, it can be retained that the emitted radiation interacts with matter.

4.3. Radiation Detection and Analysis

Radioactivity is detected and measured in the same way as every other radiation of atomic origin. Radiation detection is based on the interaction of radiation with matter.

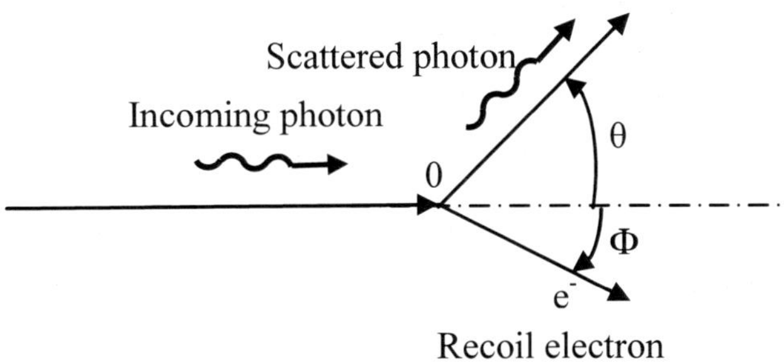

Figure 4. Scattering of a photon by a free electron (e^-), a process called Compton Effect.

4.3.1. The Interaction of Radiation with Matter

There are two forms of radiation: particle radiation and electromagnetic radiation. Particle radiation is comprised of two types: charged particle radiation and neutron radiation. The interaction of charged particles with matter is sure. That of neutrons and electromagnetic radiation is likely. In other words, the probability of interacting with matter is 100% for charged particles, and much lower than that for electromagnetic radiation and neutrons. Since the aim of this chapter is the calibration of a radiation measurement set-up, the reader is advised to consult literature on the interaction of radiation with matter for comprehensive information. In the frame of this chapter, it can just be summarized that radiation causes excitation, ionization, scattering and nuclear reactions on the atoms of matter it strikes on its way, transferring a part or the whole of its energy to them. Among the effects of the interaction of radiation with matter, scattering (Figure 4) and absorption are the most important for radioactivity assessment from measurement. Absorption denotes the transferring of the whole of the ray energy to the matter during a single interaction and it is this effect that helps to detect the ray.

According to the circumstances, gamma rays behave either like waves, or like particles. This particle character is described with the term "photon" and specifically appears in collision experiments. When a gamma-ray collides with a tiny particle of matter, it behaves like another particle and obeys the laws that govern the dynamics and the kinematics of colliding matter particles (Figure 4).

Photons have two more properties: during collisions by energies lower than 1.022 MeV, they disappear by giving up the whole of their energy to an electron that is then ejected from the atom. This photon absorption phenomenon is called "photoelectric effect". But when the photon energy is greater than 1.022 MeV, the photon disappears under emission of an electron and a positively charged "electron" named positron. This phenomenon is the pair production.

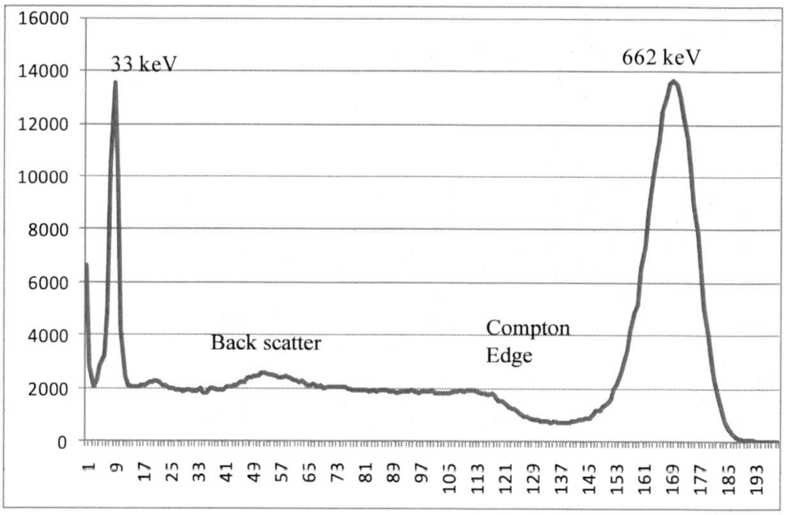

Figure 5. The gamma energy spectrum of ^{137}Cs decay (Motapon, 1990). Horizontal axis: channel numbers (energy); vertical axis: number of counts recorded in the channel considered.

4.3.2. The Energy Spectrum of a Radionuclide

Radiation *counting* is based on rays' absorption in the detector. Each ray has a given energy. A graph depicting the *count number* as a function of rays' energy, is called the energy spectrum of the sample. A radionuclide generally emits rays of definite energies. The energy spectrum of a radionuclide therefore has, according to those definite energies, one or more spectral lines or peaks that characterise that radionuclide (Figure 5) and thus enables its identification.

Following a K-shell electron vacancy from internal conversion in Barium-137, a characteristic K-X ray of Barium is emitted that appears at 33 keV on the photon spectrum.

The curve of the spectral line at 662 keV has a Gaussian shape. This means that all the gamma-rays of that line do not have the very same energy value. Their values vary around a mean value represented by the top of the peak on the spectrum. So there is an uncertainty in the energy value of the line. Energy calibration refers to the top of the peak. Energy uncertainty is tied to the shape of the peak and gives rise to energy resolution which is defined as full width of the line at half maximum (FWHM) on the spectrum. Energy resolution measures the shortest distance between two energy lines for the detection system to recognize and separate them as two different lines.

4.3.3. The Experimental Set-up

Since radioactivity cannot be felt by any of the five senses of man, physicists have developed different detection and analysis methods and corresponding experimental set-ups for each of the three types of rays. In principle, a radiation measurement set-up is comprised of a detector and the associated electronics consisting of an electric-signal preamplifier, a main amplifier, a signal discriminator and a scaler (Figure 6). A more comfortable version of the electronics is comprised of a preamplifier, a main amplifier and a multichannel analyser. Nowadays the function of the multichannel analyser is achieved by a computer equipped with a data acquisition card and appropriate software, which enables direct data processing to get the numerical results required from the measured spectrum. Other models integrate the associated electronics in a sole portable unit for outdoor radiation field measurement.

4.3.3.1. Radiation Detectors

Except for light which is detected by man's eye, any other radiation is invisible. In the course of history, physicists had to develop different methods to provide evidence for radiation [Fisher et al.]. The first of all methods, used by Becquerel, was the photographic plate: incident rays transfer their energy to AgBr molecules, each of which thereafter dissociates into atomic Ag and Br. Br evaporates whereas the dark colour of Ag provides evidence for the effect of radiation on the photographic plate. The next detector system was the gas filled capacitor. The principle here is that of incident rays ionizing gas between the capacitor plates, therefore setting electric charges free that are collected at the plates under voltage. The detector can be defined [Mann et al.] as substance in which the energy of incoming rays is converted into energy of chemical change, or into electrical, optical or thermal energy.

In the case of a gas ionization chamber, the incident ray energy transferred to an atom of the gas by a single interaction ionizes the atom, freeing two electric charges: the electron and the ionized atom or molecule. These free charges are in turn attracted and collected by appropriate electrodes in a powered electric circuit. This yields an electric signal proportional to the ray energy that released the charges. The Geiger-Müller counter works with a special version of the gas ionization chamber. Except for the photographic emulsion and dosimeters, nearly all of detector systems convert incident energy into electrical energy. Even the scintillation detector in which the incident ray energy is converted into optical energy through the fluorescence of the crystal atoms, this light nevertheless is guided onto a photocathode where it is converted into electrical energy by releasing electrons from the photocathode. The reader is advised to refer to the abundant literature on radiation detectors for more information on that matter [Siegbahn, K.]. It must just be mentioned that there are different types of detectors for the different kinds of radiation. Gas-flux detectors are used for counting beta rays; semi-conductor detectors are suitable for electromagnetic radiation and alpha particles whereas the solid scintillation detector is widely used over the World for gamma radiation which therefore is the example case in this chapter. These three types of detectors convert the incident ray energy into electrical energy by generating an electric signal in the powered circuit.

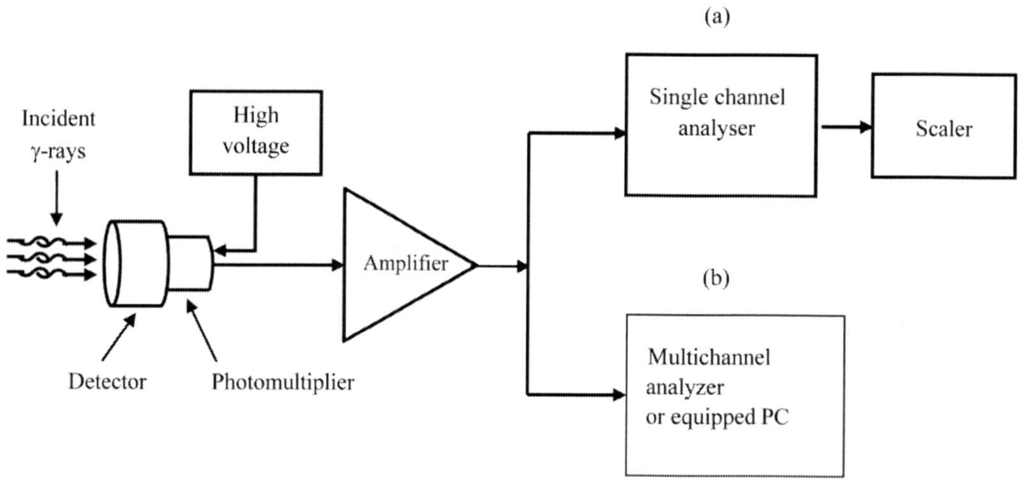

Figure 6. Radiation analysis set-up. Basic mounting with: (a) single channel analyser, and (b) multichannel analyser. The photomultiplier here acts also as preamplifier.

4.3.3.2. The Signal Preamplifier

The electric charge pulse generated in the detector must be converted into a voltage pulse, enhanced and suitably shaped to match the requirements of further signal processing by the discriminator, the scaler and the single- or the multi-channel analyser. That is the duty of the preamplifier and the main amplifier. Both components should reduce electronic noise of the circuit including the detector. Electronic noise consists of small electric pulses and competes with the pulses generated by the ray energy. It is due, for example, to random thermal movements of charged particles in the circuit.

The preamplifier is preferably charge-sensitive and converts the detector charge pulse into a voltage pulse that it enhances and preliminarily shapes, while reducing the electronic noise from detector and preamplifier material. The preamplifier output pulse is proportional to the charge generated in the detector, which itself is proportional to the ray energy deposited within the sensible volume of the detector.

4.3.3.3. The Main Amplifier

The main amplifier shapes the preamplifier output voltage pulse and boosts its amplitude by an adjustable and steady gain ranging from five to two thousand. The amplifier is voltage-sensitive and yields amplified output pulses that are directly proportional to the input pulses from the preamplifier. It is then called a linear amplifier and yields an output pulse with amplitude linear to the ray energy that induced it. Suitably shaped amplifier output pulses enable an energy analysis of the radiation spectrum by means of a signal discriminator, a single-channel analyser or a multichannel analyser.

4.3.3.4. The Signal Discriminator

The signal discriminator allows only signals with amplitudes greater than a selected threshold to pass for further processing. It is then said to work in the integral mode. It can be used to reject noise, as it rejects all pulses with amplitudes smaller than the threshold. It is important to stress that the passing pulse can be further processed, still giving information on the incident ray energy.

Most discriminator designs nowadays use a two-input (or differential-input) integrated circuit that compares the amplitude of the incoming pulse with the threshold voltage selected (Figure 7). Known as differential comparator, it can operate in a so called logic mode: it rejects all signals with amplitudes less than the threshold, which is the *logic mode* “no” or “0”, and produces a standard voltage output pulse for each signal with amplitude greater than the threshold, which is the logic mode “yes” or “1”. The produced standard or *logic* pulse is the same for all passing pulses. It no longer refers back to the incident amount of energy. It has lost its memory. It just informs that a pulse amplitude greater than the threshold has occurred. For reporting all of such pulses, that device remains an integral discriminator.

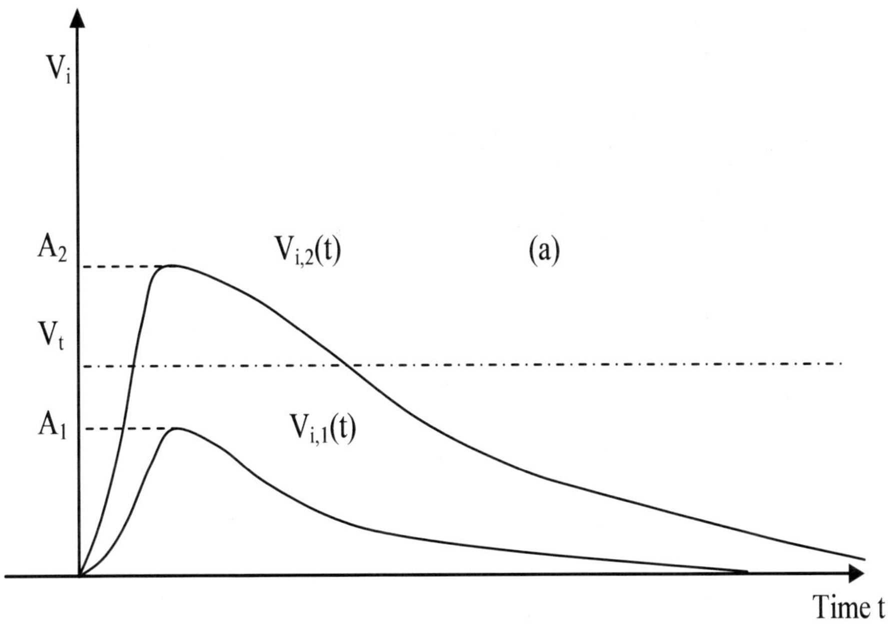

Figure 7. Continued on next page.

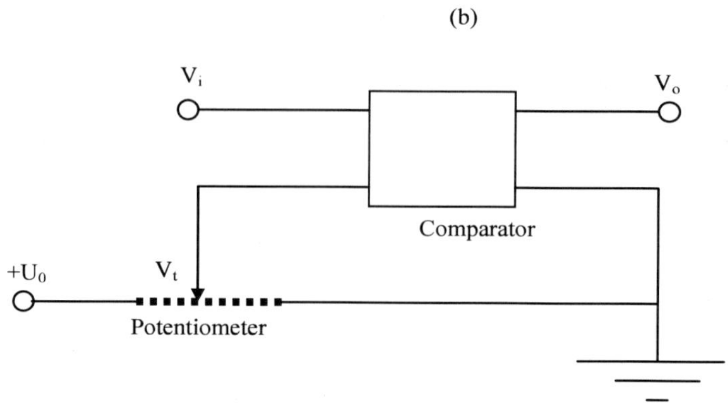

Figure 7. a) Each incoming pulse $V_{i,1}=A_1f_1(t)$ with amplitude $A_1 < V_t$ is rejected by the integral comparator circuit (b) which is then in the logic "NO" or "0" state. Each incoming pulse $V_{i,2}=A_2f_2(t)$ with amplitude $A_2 > V_t$ sets the discriminator in the logic "YES" or "1" state and therefore causes a discriminator output pulse V_0. V_t is selected on a potentiometer.

4.3.3.5. The Scaler

A scaler is an electronic component that just counts and visually displays the number of pulses or events arriving at its input within a selected interval of time. Nowadays scalers work either in the preset time mode or in the preset number of counts mode and stop counting once the preset amount is reached. They use integrated circuits operating in binary coded decimal (BCD) mode to store the number of counts and decode them back to decimal mode to display or print them.

The techniques of binary coding and decoding are very interesting. The reader is advised to eventually consult special literature [Mann et al.] for more information. The basic idea is presented below, since nuclear radiation counting is at the origin of computer operation.

The first duty of radiation counting is to tell if a ray deposited energy in the detector or not. It so gives the answer "YES" or "NO". YES corresponds to "logic 1" and NO to "logic 0". So the elementary scale of a physical counter has only two digits: 0 and 1. A counting system with only two digits is called a binary system and its digit is named "bit".

Table 1. The usual correspondence code between the decimal and the binary number systems, for decimal 0 through 10

Decimal	Binary coded			
	2^3	2^2	2^1	2^0
0				0
1				1
2			1	0
3			1	1
4		1	0	0
5		1	0	1
6		1	1	0
7		1	1	1
8	1	0	0	0
9	1	0	0	1
10	1	0	1	0

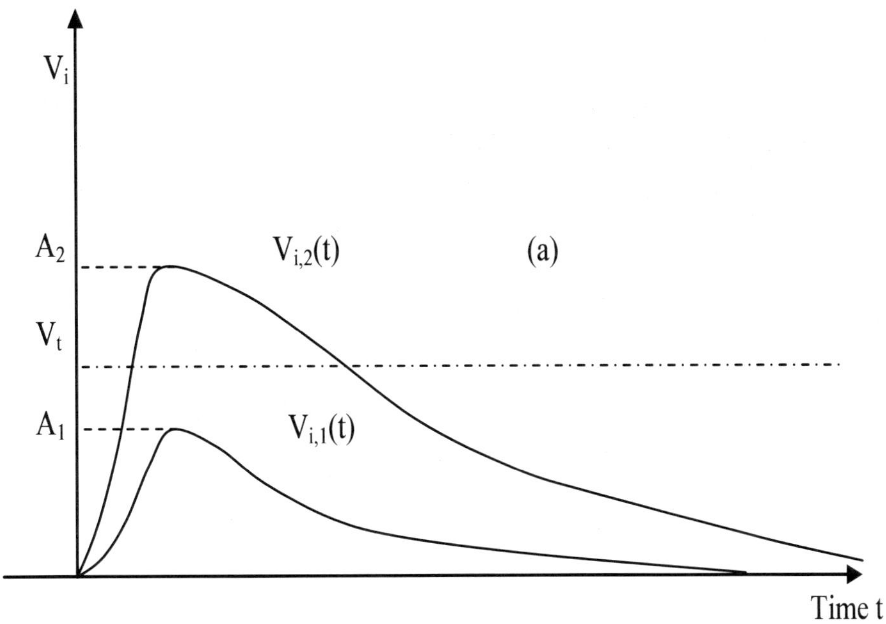

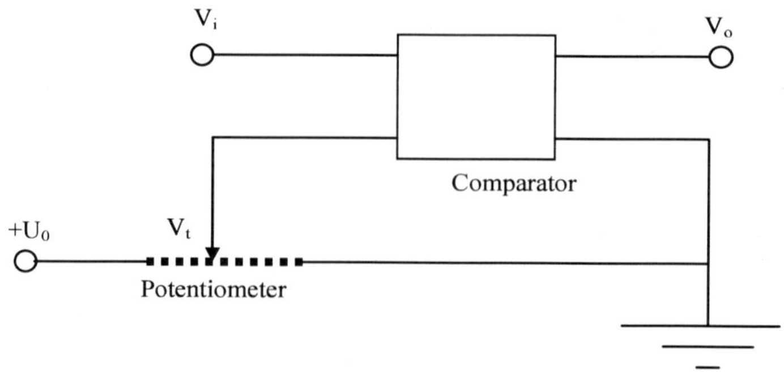

To express the number of times *logic 1* occurs in a given interval of time, one must convert binary numbers to decimal numbers; this occurs through naturally setting four bits in a series.

Table 1 presents the usual conversion coding. A series of four bits is a decade and modern scalers have a set of six decades which allows the binary codification of decimal numbers from 0 through 999999.

4.3.3.6. The Single Channel Analyser

A single channel analyser (SCA) is made of a parallel set of two discriminators and an anti-coincidence unit. The voltage threshold of each discriminator is adjustable. The threshold at one discriminator, called Lower Level Discriminator (LLD) is smaller than that of the other discriminator, therefore called Upper Level Discriminator (ULD). Each incoming pulse is entered in both discriminators at the same time. An incoming pulse of amplitude greater than the upper threshold necessarily causes both discriminators to produce, practically at the same time, an output pulse each. Both pulses synchronistically reach the two inputs of the anti-coincidence unit, which works as follows: It produces no output pulse when its inputs are reached at the same time by incoming signals. It only produces an output signal if alone the LLD sends a signal to its input, i.e. when the single channel analyser input signal has amplitude between both threshold levels.

The voltage difference between both thresholds is named "window". It can be adjusted to very narrow to allow, in combination with the scaler, operating the SCA in a way as to collect the energy spectrum data of the radiation measured, by successively increasing the lower level in steps equal to the window and measuring the count number at every step with a same duration for all steps. The spectrum in Figure 5 was measured that way, which is time- and effort consuming. Performing the energy spectrum of radiation can be achieved very comfortably with a multichannel analyser.

4.3.3.7 The Multichannel Analyser

Since the amplitude or height of the amplifier output pulse is directly proportional to the amount of energy deposited in the detector which induced it, analysing the height of such pulses provides information on the energy distribution over the rays emitted by the radiation source. The multichannel pulse-height analyser, or simply "Multichannel Analyser" (MCA), performs that duty. The aforesaid distribution corresponds to the energy spectrum of the radiation source. The MCA grades the incoming pulses according to their heights and stores the number of pulses of same height at a same address in its memory. The addresses are named "channels" or "bins" and their numbering increases linearly to the pulse-height, itself linear to the energy deposited in the detector. So a channel number corresponds to a definite energy value. Measurement with the MCA in a given time duration displays a figure depicting the count number as a function of the channel number i. e. the energy spectrum of the radiation measured.

Again the techniques used to build up a MCA are very interesting but lie outside of the scope of this chapter. Just two fundamental ideas should be presented. Firstly, at its beginning, a MCA was a series of interconnected single-channel discriminators, each of them being a channel of width ΔV equal to the aforesaid window. The whole pulse-height range of say 10V, was subdivided into $10V/(\Delta V)$ single-channel analysers. This technique is very expensive and is now abandoned. Secondly, nowadays MCAs convert the analogue pulse-height into a digital number directly proportional to it, and that pulse is *counted* in the channel whose number is equal to that digital number. The analogue-to-digital conversion is performed by charging a capacitor by means of the incoming pulse voltage, and discharging this capacitor while an oscillator of say 100MHz, is processing oscillations. The number of oscillations processed while the capacitor is discharging, is proportional to the charge and therefore to the pulse-height and defines the number of the channel in which the pulse is counted.

That process takes time, during which the MCA is blocked and can admit no other pulse. Pulses "knocking at the MCA's door" during that time are ignored. They are lost. But there exist methods to correct the counts relatively to loss caused by this *dead time*. The MCA runs in *real time*; the time it is ready to process signals is the *live time*. The difference between both is the dead time. So the counting error increases with the dead time, i.e. the higher the count rate, the higher the dead time and the higher the counting error.

Today, personal computers equipped with a data acquisition card (PCA-card) and suitable software exhibit all the features of a MCA and, furthermore, allow a large diversity and flexibility in the displaying, the manipulation, the calculation and overall exploitation of the spectrum data. Furthermore, there exist today, as previously mentioned, portable units integrating the electronics and a small MCA for outdoor radiation measurements to quickly identify radioisotopes. Nuclear electronics is a fundamental tool in radiation assessment. The International Atomic Energy Agency [I.A.E.A.] therefore provides training over this domain.

4.4. Calibrating a NaI(TL) Based Set-up for Low-level Gamma-activity Measurement in the Multi-group Approximation Method

After interconnecting and adjusting the detector and the electronic components, two different calibrations are carried out: an energy calibration and an activity calibration. Then the Multi-group approximation method is presented and, basing on total absorption and Compton spectra of gamma rays, is applied through a computer algorithm to analyzing low-level activity NaI(Tl) gamma ray spectra of such as environmental samples. A special analysis on the uncertainty estimation is included. Gamma sensitiveness is defined and numerically evaluated.

4.4.1. Experimental

4.4.1.1. Techniques and Settings

The experimental set-up (Figure 6) is comprised of a NaI(Tl) scintillation detector crystal of 76mm diameter and 76mm length [Scintillation crystal 12S12/3 PD426 Type NaI(Tl) The Harshaw Chemical Company Co. Crystal and Electronic Products. Cleveland, Ohio, USA] connected to a photomultiplier tube and *Tennelec* electronics [Tennelec 601 Oak Ridge Turnpike, P.O. Box 2560, Oak Ridge, TN 37831-2560, USA], provided with a multichannel analyser system consisting of a *Microway* 287 personal computer and an analyser card

PCA4000 (*The Nucleus*).The crystal and the photomultiplier tube (PMT) were located in a lead shield.

4.4.1.2. Energy Calibration

Energy calibration aims at identifying radionuclides from the spectrum of their characteristic ray energies. It rests upon - and permits verifying - the fact that the pulse heights are linear to the energies that induced them. The fundamental procedure consists in counting pulses from radionuclides of different ray energies and identifying the numbers of the channels in which these pulses are counted. The counting duration is the same for all measurements and complies with the activity in units of Bq, in the sense that the higher the activity, the shorter the necessary counting duration. The smallest channel number corresponds to the minimum energy value and the greatest channel number corresponds to the maximum energy value occurring from the sources facing the detector.

Energy calibration determines the relationship between the device-reading in Volts and the measurand which is the energy, so that the device-reading can be expressed in energy S.I. units of keV. It is performed by means of the linear regression. The graph depicting the channel number as a function of the ray energy, should be a line of equation $y = ax + b$, where y is the channel number i.e. the pulse height or amplitude, and x the ray energy. Slope a and y-intercept b are calculated from N measurement points $M_i(x_i, y_i)$, $1 \le i \le N$, such that σ_{xy} / S_x^2 and $b = \bar{y} - a\bar{x}$, where $\bar{x}$ and $\bar{y}$ are the mean values of the x_i and the y_i respectively, σ_{xy} the covariance of x and y, S_x^2 the variance of x and S_x the standard deviation of x, and respectively for y. When the correlation coefficient $r = \sigma_{xy} / (S_x S_y)$ is equal to or greater than $\sqrt{3}/2$, y can be assumed linear to x. One verifies that the mean point G($\bar{x}$, $\bar{y}$) belongs to the regression line of the above given equation [Courtois]. The radiation sources from the International Atomic Energy Agency used as energy standards, are listed in table 2 that presents a statistical series of the two numerical values x and y of population $N = 4$. The x values are the *energy standards*.

Table 2. Standard energies (E) and amplitudes (A) used to calibrate the lower energy part of the spectrum

i	x_i or E /MeV	y_i or A /V	Radioisotope
1	0122	1.9	Co-57
2	0.356	4.3	Ba-133
3	0.384	5	Ba-133
4	0.662	8.7	Cs-137

x is the energy deposited within the detector and y the amplitude of the pulse generated by that energy: x explains y.

From table 2, one can observe that this set of energies is suitable for radioisotope with energies not higher than that of Cs-137. Otherwise sources of higher energies up to 1.5 MeV are necessary to analyze Cs-134 or K-40. On the other hand, the modern nuclear electronics for signal processing perform a quasi-perfect linearity between pulse height and energy and the scope focuses particularly on the stability of the amplification gain. The case in table 2 is used rather as illustration to the procedure of energy calibration. An uncaliabrated MCA can always be calibrated any time by means of standard energies through assigning their values to the corresponding channel numbers read on the MCA. The linearity relationship between the energy and the channel number can be appreciated through drawing the regression line.

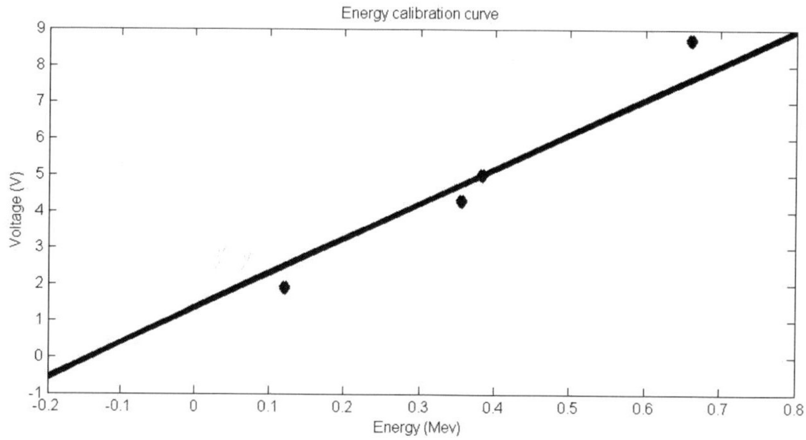

Figure 8. A Gamma energy calibration curve obtained from the data of table 2. Note: for $b>0$, electronic noise is disturbing the counting, for $b<0$, ray energies are attenuated by the detector window or any other material before reaching the active detector volume, which is the case for charged particle counting. This curve shows the limit of the regression line method.

Table 2 delivers the following data:

$\bar{x} = 0.381$, $\bar{y} = 4.975$, $S_x^2 = 0.0489$, $S_y^2 = 7.929$, $r = 0.7469 > \sqrt{3}/2$ which permits the use of a regression straight line, with $a = 9.511$ and $b = 1.351$. The equation therefore is $y = 9.511x + 1.351$. This is the explicit formulation of the relationship between the reading y on the measurement device and the standard x. The mean point $G(0.381, 4.975)$ belongs well to that straight line since $9.511 * 0.381 + 1.351 = 4.9747$ with an error term $u = |4.975 - 4.9747| = 0.0003$. Figure 8 shows the energy calibration curve obtained. The population of 4 is too small. More standard energy values are therefore quoted to get a more credible regression line.

Once this calibration line drawn, it now can be used as follows to identify the energy of rays of unknown energies: the value of the ray's channel number given by counting is mapped to the y-axis; a horizontal line is drawn at the height of that value and crosses the calibration line. Starting from this crossing point, a vertical line is traced down to the x-axis. The crossing point gives the value of the corresponding ray energy sought-after.

4.4.1.3. Activity Calibration

Activity calibration aims at estimating the activity of a radioactive substance. It rests upon the assumption that the *count rate* recorded by the counter is linear to the *activity* of the sample. The substance can contain one radionuclide, or a mixture of radionuclides. The device is calibrated with radionuclides of known activities as *activity standards*. The calibration determines the relationship between the reading of the counting device – i.e. the count rate - and the standard activity such that the reading can be expressed in Bq (Becquerel), the activity unit in the International Units System (S.I.). One Becquerel is one decay per second in the whole radioactive substance facing the detector. The relationship between the reading and the standard is described by the concept of calibration coefficient, since that coefficient turns count rates into Becquerel(s) through the calibration procedure.

(1) Calibrating the Set-up for a Sample Containing One Radionuclide

In order to assess the activity of the radionuclide of a sample containing only one radionuclide, the latter must be first identified either through its ray energy spectrum as described above, or through a calibrated MCA. The activity calibration now occurs with a source of known activity containing only that radionuclide. After the detection system to calibrate is conveniently adjusted, two spectra are measured one after the other with the same duration Δt: one for background radiation, which yields the background counts, the other for the radionuclide of known standard activity A, yielding the gross counts. Only data concerning the area or the region of the total absorption peak (TAP) are usually processed. The difference between gross and background counts gives the net counts which, when divided by Δt, yields the net count rate, say n. The measurement equation is therefore $n = S \cdot A$. Getting the linearity factor $S=n/A$ is the goal of calibrating. The activity A_p of any radionuclide of unknown activity and of measured count rate n_p is then given by $A_p = G \cdot n_p$, where $G = 1/S$; subscript p refers to "particular investigation". The uncertainty of the assessed activity will be treated later.

(2) Activity Calibration for Assessment of Radionuclide Mixtures. The Multi-Group Approximation

If the energy spectrum of the sample has shown that it contains several radioisotopes, the radiation assessment set-up must be calibrated for simultaneous measuring of those isotopes. During the measurement and due to different physical processes at stake in a radiation field, gamma rays of a given spectral line of a radionuclide are scattered and counted nearly over the whole spectrum range of the set-up. The multi-group approximation consists in sub-dividing the whole spectrum range into so called "energy groups". Energy group number i is an energy interval ranging from value E_i - ΔE_i through value $E_i + \Delta E_i$ around energy meann value E_i. In praxis, the whole spectrum range of the set-up is skilfully subdivided into as many regions, say m, as there are presumably radionuclides in the samples concerned. Photons actually belonging to the region of a definite radionuclide energy line are counted in regions of lower or, sometimes, in regions of higher energies, polluting the background of these regions. The contribution of isotope k of activity A_k to the count rate n_i of region i is taken into account through the scattering coefficient s_{ik}. So that the total count rate n_i results from the contributions of the activities of all radioisotopes of the sample:

$$n_i = \sum_k s_{ik} A_k \, . \tag{1a}$$

For a mathematical description, the coefficients s_{ik} are the elements of a (m, m)-matrix, say $S(s_{ik})$, n_i and A_k are elements of column- or, respectively, line-matrices, say $N(n_i)$ and $A(A_k)$. So the matrix equation for eq. (1a) is

$$N = S \cdot A. \tag{1b}$$

For calibration purpose, a one-radionuclide source of known activity must be prepared for each of the radionuclides of the radionuclide mixture sample. The elements s_{ik} are then determined in presence of radioisotope k of activity A_k for net count rate n_i in region i in the absence of all other radioisotopes, so that its measurement equation is

$$s_{ik} = \frac{n_i}{A_k}. \tag{2}$$

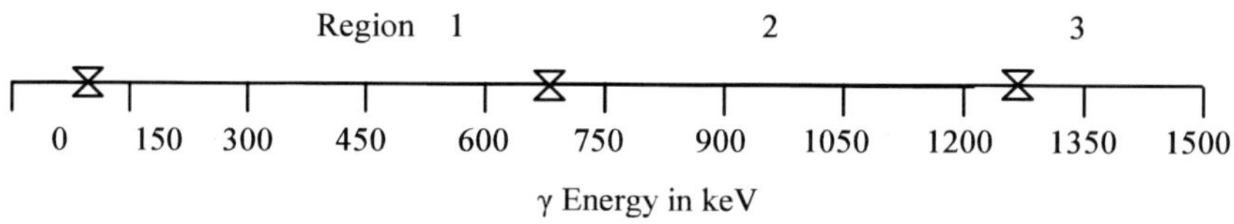

Figure 9. Spectrum subdivision into three regions numbered 1 through 3. Symbol ⧖ denotes region limits. The setting of the region limits obeys to the effective energy resolution that depends on the shape of the peaks.

The energy axis of the set-up was calibrated with standard gamma energies of radioactive sources, namely ^{22}Na with its 511, 1275, and 1786 keV lines, ^{137}Cs with its 662 keV line and finally a uranium-radium geological sample with energy lines at 185, 242 and 395 keV. But, as aforesaid concerning the Gaußian shape of the energy lines on a measured spectrum, each line is broad and spreads over an energy interval around a mean value. That interval is the energy group, of the given mean value. Photon groups on a measured spectrum are usually called regions and set following single lines.

The particularity in the Multi-group approximation is now that “energy calibration” actually assigns energy groups to regions of the MCA spectrum. Taking into account the energy lines of the searched radioisotopes ^{134}Cs, ^{137}Cs and ^{40}K, the region boundaries in the case study of this chapter were staked as follows (Figure 9):

Region 1: spreading from channel 148 (corresponding to 110 keV) to channel 1037 (678 keV), assigned to the group of energy mean value 662-keV of the line of ^{137}Cs;

Region 2: from channel 1039 (680 keV) to channel 1874 (1256 keV), assigned to the group of the 796-keV line of ^{134}Cs;

Region 3: from channel 1876 (1258 keV) to 2040 (1620 keV), assigned to the group of the 1461-keV line of ^{40}K.

A spectrum of each isotope was measured during 54000 seconds and effective scattering coefficients s_{ik} were determined column by column following eq. (2).

The activity calibration coefficients were deduced from scattering coefficients determined by means of standard activities of sources filled in 250 cm^3 Marinelli cups, namely two liquid sources of ^{137}Cs, ^{134}Cs of activity concentrations of 3.78 and 2.55 Bq g^{-1} respectively and a solid pure KCl compound. All of them had been calibrated in the Nuclear

Physics Laboratory of the Faculty of Science in Rabat, Morocco, which is a reference laboratory associated to the *I.A.E.A*.

4.4.1.4. Uncertainty Estimates of the Assessed Activities A_i

The assessed activities A_k are solutions of the matrix equation $A = S^{-1}N$. The elements n_i of matrix *N* here are the count rates determined in each region *i* by one measurement of the energy spectrum of the sample. S^{-1} is the inverse matrix to *S*, say $G(g_{ki})$; so that $A = GN$ and the equation for the respective elements

$$A_k = \sum_{i=1}^{m} g_{ki} n_i \tag{3}$$

yields the assessed activities from the measured count rates n_i.

The relative uncertainty of the scattering coefficient s_{ik} is [Taylor and Kuyatt, 1993]:

$$u_r^2(s_{ik}) = u_r^2(A_k) + s_r^2(n_i), \tag{4}$$

where $u_r(A_k) = \dfrac{u(A_k)}{A_k}$ and $s_r(n_i) = \dfrac{s(n_i)}{n_i}$. $s(x)$ or $u(x)$ is the uncertainty of the quantity x. Index *r* means relative uncertainty. Symbol ***s*** applies to uncertainty evaluated by a **s**tatistical analysis of series of observations. Symbol ***u*** applies to **u**ncertainty evaluated by means other than statistical analysis of series of observations as shown in the following example.

(1) The Uncertainty of the Calibration Coefficients

From the basic expression $s_{ik} \pm u(s_{ik})$, the limits $s_{ik,\min}$, $s_{ik,\max}$ and respectively corresponding $g_{i\ell,\max}$ and $g_{i\ell,\min}$ were computed, then the uncertainty

$$u(g_{i\ell}) = \frac{1}{2}\left|g_{i\ell,\max} - g_{i\ell,\min}\right| \text{ and finally } u_r(g_{i\ell}) = \frac{u(g_{i\ell})}{g_{i\ell}}. \tag{5}$$

(2) The Uncertainty of the Activities

To estimate the uncertainty $u(A_i)$ of the activity A_i given in eq. (3), one may consider variable $g_{i\ell}$ of uncertainty $u(g_{i\ell})$ and relative uncertainty $u_r(g_{i\ell})$. Te mathematical point of view

$$dA_i = \sum_{\ell=1}^{3}\left(\frac{\partial A_i}{\partial g_{il}} dg_{i\ell} + \frac{\partial A_i}{\partial n_\ell} dn_\ell\right) \text{ over the form}$$

$$u(A_i) = \sum_{\ell=1}^{3} \left(n_\ell \, u(g_{i\ell}) + g_{i\ell} \, s(n_\ell) \right) \tag{6a}$$

has the statistical shape

$$u^2(A_i) = \sum_{\ell=1}^{3} \left(n_\ell^2 u^2(g_{i\ell}) + g_{i\ell}^2 s^2(n_\ell) \right) . \tag{6b}$$

The relative uncertainty $u_r(A_i)$ to multiply with the mean value A_i for estimating the uncertainty $u(A_i)$ itself is assessed through

$$u_r^2(A_i) = \sum_{\ell=1}^{3} \left(u_r^2(g_{i\ell}) + s_r^2(n_\ell) \right). \tag{6c}$$

4.4.2. Results

4.4.2.1. Effective Scattering Coefficients

(1) Calibration Input Data

Table 3 shows the calibration data: the activities of radioisotopes and their absolute uncertainties, gross counts and background counts in the three regions. The same background counts were used in the same region for all of three isotopes. Gross counts in region 3 in the presence of ^{137}Cs alone are lower than background.

Table 3. Calibration activities, their uncertainties and corresponding gross counts in the three regions

Observation	1	2	3	4
Isotope	^{137}Cs	^{134}Cs	^{40}K	Background
Activity/Bq	**886**	**548**	**3416**	
Uncertainty/Bq	**21**	**13**	**170**	
Region 1	6 308 154	6 946 000	1 859 199	488 403
Region 2	373 545	1 968 171	820 116	140 709
Region 3	48 226	229 040	663 340	48 617

Counting duration t = 54000 seconds. Background counts in the last column.

According to Table 3a, ^{137}Cs contributes to counts of region 2, as does ^{134}Cs to those of region 3. The ^{134}Cs is counted to 76.28 % in region 1 and ^{40}K is counted to 25.53 % in region 2 and to 51.42 % in region 1. This makes together 77 % in regions of clearly lower energies. The counts of the three radioisotopes in the three regions yield a 3x3 counting matrix. The ^{137}Cs and ^{134}Cs contribute in the same order of magnitude (4.31 and 3.34) and with the same uncertainty (0.01) to counting of the respectively higher neighbour region. The count rate of ^{137}Cs in region 3 is 0.

Spectrum subdividing is such that two energy lines (569.32 and 604.71 keV) of ^{134}Cs lie in region 1, and two other (795.87 and 801.95 keV) in region 2. Over 100 decays (Browne and Firestone, 1986), the photon number ratio between the two lines of region 1 and the two lines of region 2 should be 1.2. But here it is 3.53. This expresses an intensive scattering from region 2 to region 1, noticeably raising ^{134}Cs counts in region 1 compared to region 2. The same is true for ^{40}K with its line in region 3, where one notices that the contribution to counting in lower region (1 and 2) rises with distance from region 3.

The counts of ^{137}Cs in region 3 and observation 1 are lower than those of the background. Such an effect could come from statistical fluctuation, but it could also be systematic as a consequence of the dead time.

(2) Scattering Coefficients

The quantity $\sum_i s_{ik}$ of suggested symbol S_k is the total net count rate per second and Becquerel of activity of isotope *k*. It measures the sensitiveness of the detection system to one Becquerel activity of the radioisotope concerned. Counting occurs simultaneously in the three regions in the presence of one isotope at the same time: either ^{137}Cs as column 1, or ^{134}Cs as column 2, or ^{40}K as column 3. Table 4b shows that the net counting rate per Bq varies from an isotope to another. Concerning the three isotopes, net counting rate per Bq is maximal for ^{134}Cs with value 286.08 and minimal with value 14.45 for ^{40}K that however has the highest activity of the three isotopes, and also the highest energy (1 460 keV). This is partly due to the photon absorption cross section which decreases when the energy rises and is therefore minimal for ^{40}K, and partly to the fact that a large number of photons is scattered out of the detector by multiple scattering which brings them to lower energies, so that relatively few thermalized photons related to the ^{40}K activity reach the detector.

The ^{137}Cs and ^{134}Cs contribute to counting of the respectively higher neighbour region. The ^{137}Cs contributes through random pile-up, whereas ^{134}Cs may contribute to counting in region 3 through random pile-up and coincidence of its two cascades (569.32 $\Rightarrow$ 604.71 and 795.87 $\Rightarrow$604.71), its contribution in region 3 (6.10) is therefore greater than that of ^{137}Cs in region 2 (4.87). Similarly, 77% of ^{40}K photons are counted in the two regions of clearly lower energies, reflecting the behavior of the cross section of the photoelectric effect, which rises when photon energy decreases. Photoelectric effect is the basis of NaI spectrometry.

4.4.2.2. Calibration Coefficients

Values of the calibration coefficients $g_{i\ell}$ (table 5a) vary from -39.18 to +339.89. This is a relatively large range of remarkable magnitude (10^2). The negative sign of all elements outside a diagonal comes from multiplication by $(-1)^{i+\ell}$ and yields perfectly symmetrical signs relatively to each diagonal.

Table 4. Calibration data: (a) Net count rates of isotope *k* in region *i* are in s^{-1} with counting error (Poisson) in brackets as combined absolute uncertainty from gross and background count rates. (b) Scattering coefficient s_{ik} of isotope *k* towards region *i* are in units of 10^{-3} counts per second and Becquerel and (c) their relative uncertainties in percents. Index *i* is number of region, *k* that of radioisotope

(a)

Observation	1	2	3
Isotope **Region**	^{137}Cs	^{134}Cs	^{40}K
1	107.77 (0.05)	119.58 (0.05)	25.38 (0.028)
2	4.31 (0.01)	33.84 (0.027)	12.60 (0.018)
3	-0.0072 (0.006)	3.34 (0.01)	11.38 (0.016)

(b)

Region	^{137}Cs	^{134}Cs	^{40}K
1	121.6 (2.9)	218.22 (5.18)	7.43 (0.37)
2	4.87 (0.12)	61.76 (1.47)	3.69 (0.18)
3	0.00 (0.00)	6.10 (0.15)	3.33 (0.17)
$\sum_{i=1}^{3} s_{ik}$	126.47	286.08	14.45
Activity/Bq	886	548	3416

(c)

i \ k	^{137}Cs	^{134}Cs	^{40}K
1	2.37	2.37	4.98
2	2.39	2.37	4.98
3	79.63	2.30	4.98

Table 5. Calibration coefficients and test results. (a) Calibration coefficients g_{it} are in units of [$s \cdot Bq$]. (b) Test result activities are in Bq. Each sample was observed one time (*i*) and contained only one isotope (*k*)

(a)

Isotope	Region 1	Region 2	Region 3
^{137}Cs	9.66	-35.93	18.24
^{134}Cs	-0.86	21.36	-21.74
^{40}K	1.59	-39.18	339.89

(b)

Observation	^{137}Cs	^{134}Cs	^{40}K
1	**886.**0007	$- 8.032 \times 10^{-4}$	1.26×10^{-2}
2	$- 6.77 \times 10^{-5}$	**548**	-7.93×10^{-5}
3	0.66	-0.396	**3416.**725

All elements on the main and the secondary diagonals are positive, whereas all elements outside the diagonals are negative. This reminds of the situation in electrostatics where all elements with ($i \neq k$) are negative and called coefficients of influence and all elements on the main diagonal (i=k) are positive and called the 'capacity' of the conductor i. Similar to the conductors, radioisotopes of a sample mutually influence the counting of the activity of each other.

For validation of the computing algorithm designed in this work to assess activities of radioisotopes of an environmental sample and their respective uncertainties, the algorithm was run with the calibration counting data to compute activities and uncertainties and compare them with known values. Table 5 shows the results.

Ordered in a series, the three observations build a matrix of same order like the calibration coefficient matrix and with same signs of elements. This enables computing three observations at one run.

The computed activities are located on the main diagonal (bold) and are practically the same as the known calibration activities. The main diagonal therefore has a determinant role. Eventual deviation is thoroughly negligible compared to the known uncertainty. So it can be concluded that the algorithm is reliable and matches the task. Since each observation contains only one radioisotope, the minus sign should denote activity 0.

4.4.2.3. Special Analysis on the Activity Uncertainty

The uncertainty measures the dispersion of the values that could reasonably be attributed to the measurand. This is demanded for calibration according to ISO 17025 requirements [Template.cfm].

(1) The Uncertainty of the Scattering Coefficients

Table 4c shows computed relative uncertainties of the scattering coefficient s_{ik}. The relative uncertainty for the contribution of ^{137}Cs and ^{134}Cs is about 2.4 % except for the contribution of ^{137}Cs to region 3 where it is 79.63 % because the contribution of ^{137}Cs in that region is 0; the value 79.63 % may just express background fluctuation due to deficient statistics, and surely dead time effect [Owono].

(2) The Uncertainty of the Computed Activities in the Multi-group Approximation

The uncertainty $u(A_i)$ of the activity A_i may be estimated through Eqs. (6a, b and c) above. Table 6a shows the results. Only Eq. (6a) gives for ^{134}Cs in observation 2 an uncertainty value of 13.56, very close to that of the calibration activity, 13. Eq. (6b) with 9.59 is also rather close to 13. So even for the three radioisotopes, the mathematical approximation (6a) shows the best approach here; its statistical expression (6b) behaves alike. Both equations yield results with a somehow acceptable order of magnitude for want of anything better. Eq. (6c) must be eliminated since it systematically yields very high values. Definitely none of the three equations fully agrees with known uncertainties of the calibration activities. Further investigation is thus necessary, which leads to the solution formula

$$u_r^2(A_i) = \sum_{\ell=1}^{3}\left(u_r^2(s_{\ell i}) + s_r^2(n_\ell)\right). \qquad (6c')$$

Table 6. Uncertainty values of the calibration activity. Uncertainties known (bold)and computed from: (a) eqns.6a, b, c; (b) modified eqn. 6c'.

(a)

Isotope	^{137}Cs	^{134}Cs	^{40}K
Observation	[Bq]	[Bq]	[Bq]
1	**21** 6a = 11.26 6b = 9.69 6c = 717.47		
2		**13** a = 13.56 b = 9.59 c = 77.70	
3			**170** a = 118.36 b = 89.43 c = 501.84

(b)

Observation	^{137}Cs	^{134}Cs	^{40}K
1	**21** 6c' = 21.01		
2		**13** 6c' = 13.01	
3			**170** 6c' = 170.17

Table 7. Activity uncertainty matrix computed through counting data sequences 1-2-3, 2-3-1 and 3-1-2, in Bq. Computation by means of: (a) eqn. 6c', (b) transposed relative uncertainties of the scattering coefficients.

(a)

Observation	^{137}Cs	^{134}Cs	^{40}K
1	**(21)** 21.01	13.10	2720.57
2	21.03	**(13)** 13.01	81.80
3	44.10	27.29	**(170)** 170.17

(b)

Observation	^{137}Cs	^{134}Cs	^{40}K
1	**(21)** 21.01	13.01	170.14
2	21.18	**(13)** 13.01	170.17
3	705.48	13.10	**(170)** 170.16

Table 8. Corrected uncertainty results computed by means of transposed relative uncertainties of the scattering coefficients according to last formula (6c') above, in Bq

Observation	^{137}Cs	^{134}Cs	^{40}K
1	**(21)** 21.00	13.01	170.10
2	21.00	**(13)** 13.01	170.10
3	21.00	13.01	**(170)** 170.10

Table 8 shows the results.

- Each observation yields the very same results and is therefore reliable.
- Furthermore, to avoid confusion, let the calibration activity A_k and count rate n_i be denoted A'_k and respectively n'_i. Then the relative uncertainty of the calibration data is described by $u_r^2(s_{\ell i}) = u_r^2(A_i') + s_r^2(n_\ell')$ and that of the assessed activities by
- $$u_r^2(A_i) = \sum_{\ell=1}^{3}\left(u_r^2(A_i') + s_r^2(n_\ell') + s_r^2(n_\ell)\right). \quad (7)$$

  This demonstrates how the uncertainty of the assessed activities really depends upon that of the calibration data, although the way to this discovery was intuition. But a similar result was systematically obtained afterwards for the simple case of a one-radioisotope sample.

(3) The Uncertainty of the Assessed Activity of a Sample With one Radioisotope

Coming back to the uncertainty of the assessed activity of a sample with only one radioisotope, let *G=1/S* be the calibration coefficient of relative uncertainty given through $u_r^2(G) = u_r^2(A') + s_r^2(n')$, which has proved to be the best fitting formula for error propagation in activity analysis [Owono], where *A'* and *n'* are the calibrating activity and count rate respectively. The activity A_p of any radionuclide of unknown activity and of measured count rate n_p is then $A_p = G \cdot n_p$ of relative uncertainty given by

$$u_r^2(A_p) = u_r^2(G) + s_r^2(n_p) \text{ or } u_r^2(A_p) = u_r^2(A') + s_r^2(n') + s_r^2(n_p), \quad (8)$$

which turns out to be the basic structure of Eq. (7) that it therefore validates. So the uncertainty of the assessed activity is specifically tied to that of the calibrating data and cannot be better than that.

(4) The Consequence

Classical theory of statistics states that

$$u_r^2(A_i) = \sum_{\ell=1}^{3} \left(u_r^2(g_{i\ell}) + s_r^2(n_\ell) \right). \tag{9}$$

But a practical solution turns out to be

$$u_r^2(A_i) = \sum_{\ell=1}^{3} \left(u_r^2(s_{\ell i}) + s_r^2(n_\ell) \right), \tag{10}$$

which evidently is a universal formula, since it applies to the calculation of errors of the solutions of any system of linear equations of m unknowns when the errors of the elements of the coefficient matrix are known. Another major conclusion is that

$$u_r^2(g_{i\ell}) = u_r^2(s_{\ell i}) . \tag{11}$$

$u_r^2(g_{i\ell})$ are the m^2 elements of the relative error matrix to the calibration coefficients' matrix which is defined as the inverse matrix to the scattering coefficients' matrix. $u_r^2(s_{\ell i})$ are the m^2 elements of the relative error matrix to the scattering coefficients' matrix.

It therefore can be stated that the relative-error matrix to the inverse matrix is the transposed to the relative-error matrix to the initial matrix. Simpler: 'In matrix terms, the relative error of the inverse matrix is the transposed to the relative error of the initial matrix.'

CONCLUSION

The NaI(Tl) scintillation detector is still topical the World over. Efforts to improve its reliability for gamma spectroscopy are going on in many laboratories up to developing new spectrometric analysis methods [Muminov et al., Vlachos].

The multi-group approximation using scattering or detection coefficients and applied to NaI(Tl)-gamma spectrometry device, raises its gamma ray sensitiveness, enables better computing of the uncertainties and yields more reliable analysis results for low-level activity environmental samples. As must be stressed, this method sets in evidence two important achievements: the first one is intuitively discovering a universal matrix formula for the assessment of the uncertainty of measured activities and, fortunately, a formula that applies to the calculation of errors of the solutions of any system of linear equations of m unknowns when the errors of the matrix elements are known. The second important achievement is the evidence, in matrix terms, that the relative error of the inverse matrix is the transposed matrix to the relative error of the original matrix.

REFERENCES

Browne, E., and Firestone, R.R. Table of radioactive isotopes (Wiley and Sons, New York, 1986).

Courtois, C. Dossier n° 9: Exemples de Traitement d'une série statistique à deux variables numériques, 2003. http://megamaths.perso.neuf.fr/ oral2/csta-03828.pdf (accessed 25.10.12).

Fisher, C., editor. *Les radioéléments et leurs utilisations.* (Eyrolles, Paris, 1980).

I.A.E.A. Selected Topics in Nuclear Electronics. IAEA-TECDOC-363, (Vienna 1986).

Mann, W.B., Ayres, R.L., Garfinkel, S.B. Radioactivity and its Measurement, 2nd ed. (Pergamon Press, Oxford, 1980).

Motapon O. Conversion Interne: cas du 56Ba137. *Mémoire de Maîtrise de Physique,* Université de Yaoundé, 1990 (unpublished).

Muminov, I. T., Muhamedov, A. K., Osmanov, B. S., Safarov, A. A., Safarov, A. N. Application of NaI(Tl) detector for measurement of natural radionuclides and (137)Cs in environmental samples: new approach by decomposition of measured spectrum. J. *Environ. Radioact.* 84(3), 321-331 (2005).

Owono, A. P. Multi-group approximation, scattering and calibration coefficients, uncertainty estimates and detection limits of a NaI(Tl)-based gamma spectrometry set-up for low-level activity analysis. J. *Environ. Radioact.* 101(4), 692-699 (2010).

Payne, G. C. Calibration: What Is It? *Quality Progress.* 2005 (May). http://asq.org/quality-progress/2005/05/measure-for-measure/calibration-what-is-it.html (accessed 10.11.12).

Siegbahn, K. Editor. Alpha-, Beta- and Gamma-ray Spectroscopy. 5th print. (North-Holland Publishing Company, Amsterdam, 1979).

Taylor, B. N. and Kuyatt, C. E. Guidelines for Evaluating and Expressing the Uncertainty of NIST Measurement Results. NIST Technical Note 1297 (1993). http://physics.nist.gov/cuu/Uncertainty/basic.html. (accessed 22.04.09)

Template.cfm (Objet application/pdf). Calibration Principles. http://www. isa.org/Template.cfm?Section=Find_Books1&template=Ecommerce/FileDisplay.cfm&ProductID=7577&file=ACFBA59.pdf (accessed 11.11.12).

University of Toronto, Module l6: Radiation Detection and Measurements. http://www.ehs.utoronto.ca/Programs_and_Services:RadiationSafety:radtraining:mo(accessed 10.11.06).

Vlachos, D. S. Self-calibration techniques of underwater gamma ray spectrometers. *J. Environ. Radioact.* 82(1), 21-32 (2005).

In: Calibration Technology, Theories and Applications ISBN: 978-1-62618-808-2
Editors: Ikumatsu Fujimoto and Kunitoshi Nishimura

Chapter 5

A HYBRID SCE-UA-KNN OPTIMISATION ALGORITHM APPLIED TO THE CALIBRATION OF A RAINFALL-RUNOFF MODEL

Hamouda Dakhlaoui*[1] *and Zoubeida Bargaoui
Université Tunis El Manar, Ecole Nationale d'Ingénieurs de Tunis,
Tunis, Tunisia

ABSTRACT

Rainfall-Runoff modeling is of crucial importance in water resources management. Effectively, it aims to predict river runoff fluxes as a response to rainfall, air temperature and solar radiation time series for given areas represented by river watersheds. Most Rainfall-Runoff models are based on the conceptualization of the dynamics of the water balance over the watershed domain (hillslopes and hydrographic network) for given soil and topography as well as vegetation information. Mass conservation and momentum equations are solved to predict runoff outputs. Parameters of rainfall-runoff models are generally calibrated using river runoff observations through the adoption of adequate objective functions so as to minimize the variance of model errors. The calibration of Rainfall-Runoff models presents several difficulties such as (i) interdependence between model parameters, (ii) indifference of objective function to some parameters, (iii) difficulty of definition of gradient direction, (iv) existence of local optimums and (v) scale problems related to parameters .The Shuffled Complex Evolution of the University of Arizona (SCE-UA) (Duan et al. 1993) global optimisation method is recognized as the most effective optimisation algorithm applied to achieve rainfall runoff model calibration. However, this algorithm could be time consuming, especially for the case where several model parameters are to be fitted or when little modeling time or space steps are adopted. The purpose of this study is to improve the efficiency of this optimization algorithm. Calibration efficiency improvement is performed using the k-Nearest Neighbors (KNN) approach in estimating the objective function. The motivation is that, when objective function is evaluated by model simulation, the computation time may be too long. Now, a database of both explored parameters and corresponding

[1] Corresponding Author: E-mail: hammouda.dakhlaoui@laposte.net

objective function values is constituted as SCE-UA is running. In order to save computing time, it is therefore proposed to take advantage of this database by estimating the objective function indirectly using an interpolation by the means of nearest neighbors in the parameter space. The problem of how to choose the situations where objective function is estimated by KNN rather than by model simulation is addressed here. We propose an enhanced algorithm that makes a choice based on the ranking of the explored population by their objective function values. The HBV rainfall runoff model (Begström, 1976) of the Swedish Meteorological and hydrological Institute (SMHI) is adopted as rainfall – runoff model. Provided observations of rainfall and resulting river discharge at the watershed outlet, the model parameters are calibrated using an objective function based on prediction error statistics. Thus, evaluation of the model performance is performed through the multi objective criterion RV = Nash–ω|RD|, where Nash is Nash-Sutcliffe criteria (Nash and Sutcliffe, 1970) which is a standardised form of runoff squared error, ω weight and RD runoff relative bias. The effectiveness of this hybrid optimization algorithm is tested for synthetic runoff data derived from Sejnène watershed (Tunisia). Analysis of the results suggest that an improvement on efficiency of about 40% is obtained.

Keywords: KNN, Model calibration, Shuffled Complex Evolution, Optimisation, Rainfall runoff model

1. Introduction

Rainfall-Runoff models are used widely in different fields such as water quantity and quality management, flood forecasting, and climatic changes studies. The successful use of these models depends on the quality of the estimation of their parameters. Generally, the calibration of such models could be considered as an optimisation problem that maximises the agreement between observed and simulated historical runoff. This optimisation problem is well known for carrying some difficulties such as interdependence between model parameters, the indifference of objective function to some parameters, difficulty of definition of gradient direction, the existence of local optimums and scale problems related to parameters.

The stochastic global optimisation methods such as genetic algorithm (Goldberg, 1989), Shuffled Complex Evolution (Duan et al., 1993) or simplex simulated annealing algorithm (Bates, 1994) are found to be the most powerful methods to overcomes such optimisation difficulties and to give best results with calibration of rainfall-runoff models (Wang, 1991, Duan et al., 1993, Efstratiadis and Koutsoyiannis 2002; Franchini, 1996; Franchini et al., 1998; Gupta et al., 1999; Thyer et al., 1999).

In this direction, several researches (Kuczera (1997); Franchini et al. (1998); Thyer et al. (1999)) proved that the optimization algorithm Shuffled Complex Evolution of the University of Arizona (SCE-UA, Duan et al., 1993) is one of the most effective optimization algorithms when applied to the calibration of Rainfall Runoff models, compared to other optimization algorithms such as Genetic algorithm, Pattern search coupled with Sequential Quadratic Programming, Simulated Annealing Simplex algorithm. This is so especially for the most complicate cases (such as distributed model when both parameters and inputs are assumed space varying).

Despite the fact that it is efficient and effective, the SCE-UA is time consuming when applied to the calibration of rainfall-runoff models. In fact it needs a great number of objective function evaluations to reach the optimal solution (Tolson and Shoemaker, 2007) especially in the case of a huge number of parameters to be calibrated (e.g. the case of semi-distributed model where parameter values have to be varied with sub-catchments). This problem becomes more and more annoying particularly when one's estimation of the objective function is time expensive. Such situation occurs when the calibration period is long or when time step modelling is short. This difficulty could restrict the use of SCE-UA for some applications such as model parameter updating or hydrologic forecasting which is a drawback.

Since the objective function estimation by model simulation is the most time consumer in the optimisation process, the reduction in the number of evaluations of objective function needed by calibration algorithm has been proposed as an important way to improve its efficiency. In fact, the implementation of the approximation functions is one of the main techniques adopted for this purpose.

This method involves estimating the objective function from the data sets of parameters explored by the optimization algorithm instead of estimating by direct simulation using the rainfall-runoff model. The approximation methods proposed in the literature are polynomial interpolation, geostatistical interpolation, the K-Nearest Neighbours method, the Radial Basis functions. Such techniques have been used in evolutionary algorithms for the optimization of expensive functions in computing time (Ong et al. 2003, Ostfeld and Salomons 2005, Zhou et al. 2007). However, applying function approximations may affect the ability of the optimisation algorithm to find the optimal solution.

In effect, the estimating error of the approximation function could perturb the algorithm and direct its search mechanism in the wrong direction. Thus, the introduction of approximation function reduces the computation time but at the expense of the estimating the objective function accuracy. To reduce the number of simulations required for convergence without affecting the ability of the algorithm to identify the optimal solution represents a challenge. It is assumed that a compromise can be found for estimating the objective function by using the simulation model or the approximation function. Several approaches have been proposed to "optimize" the use of approximation functions, and they have always been adapted to the optimization algorithm used (Giannakoglou 2002, Ong et al. 2003, Ostfeld and Salomons 2005, Zhou et al. 2007, Dakhlaoui et al. 2012). In this study, it is proposed to adopt the K-Nearest Neighbours (KNN) to approximate the objective function.

The remainder of the paper is organized as follows. In section 5.2 the techniques and tools used are discussed: SCE-UA optimisation algorithm, and HBV rainfall-runoff model structure, the chosen approaches of model calibration, the objective function, and the studied catchment's data. Section 5.3 recalls the K-Nearest Neighbours technique (KNN) and the method of its implementation to SCE-UA. In the last section, we present the results of the application of KNN technique to the calibration of the HBV hydrological model to the studied basin.

2. Methods and Tools

The general aim is to investigate the implementation of SCE-UA in HBV model to solve the optimisation problem ensuing from its calibration. It is proposed to adopt the K-Nearest Neighbours as an approximation function of the objective function in the SCE-UA algorithm. We focus on its effect on the efficiency and effectiveness of the optimization algorithm. We first present the basis of the rainfall-runoff model.

2.1. HBV Rainfall Runoff Model

A Rainfall-runoff model is a mathematical model describing the rainfall-runoff relations with a catchment scale. The main components of a rainfall runoff model are described in Figure 1. The model reproduces time series of the surface runoff hydrograph, evapotranspiration and groundwater recharge (as outputs) as a response to a rainfall hyetograph and air temperature series (as inputs). In the present case we focus on the surface runoff hydrograph as output of interest and we evaluate model performance with respect to runoff data.

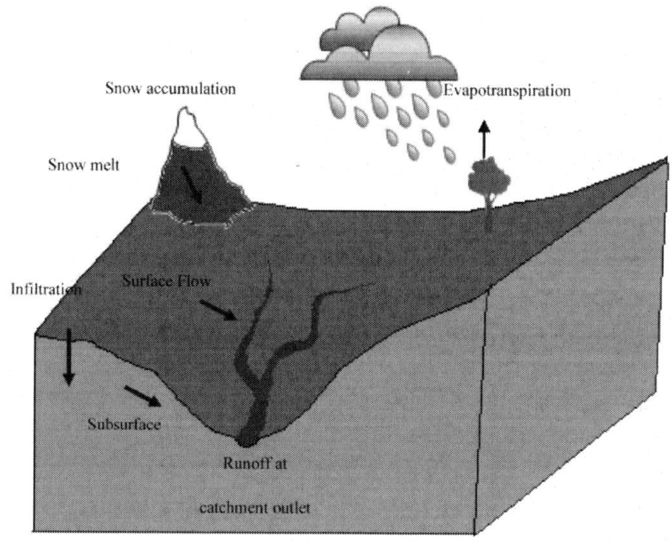

Figure 1. The main component of HBV Rainfall-Runoff model.

Rainfall-Runoff models have found successful applications to different fields such as hydrological series reconstitution, water quantity and quality management, flood forecasting, climatic change studies, etc., especially conceptual model based on the representation of soil horizons as linear or non-linear reservoirs.

The HBV model is one of the most successful conceptual model. It's of Swedish origin (Begström, 1976). It has been applied in more than 30 countries. It's performance has been proven under different climatic conditions (Lindstrom et al. 1997).

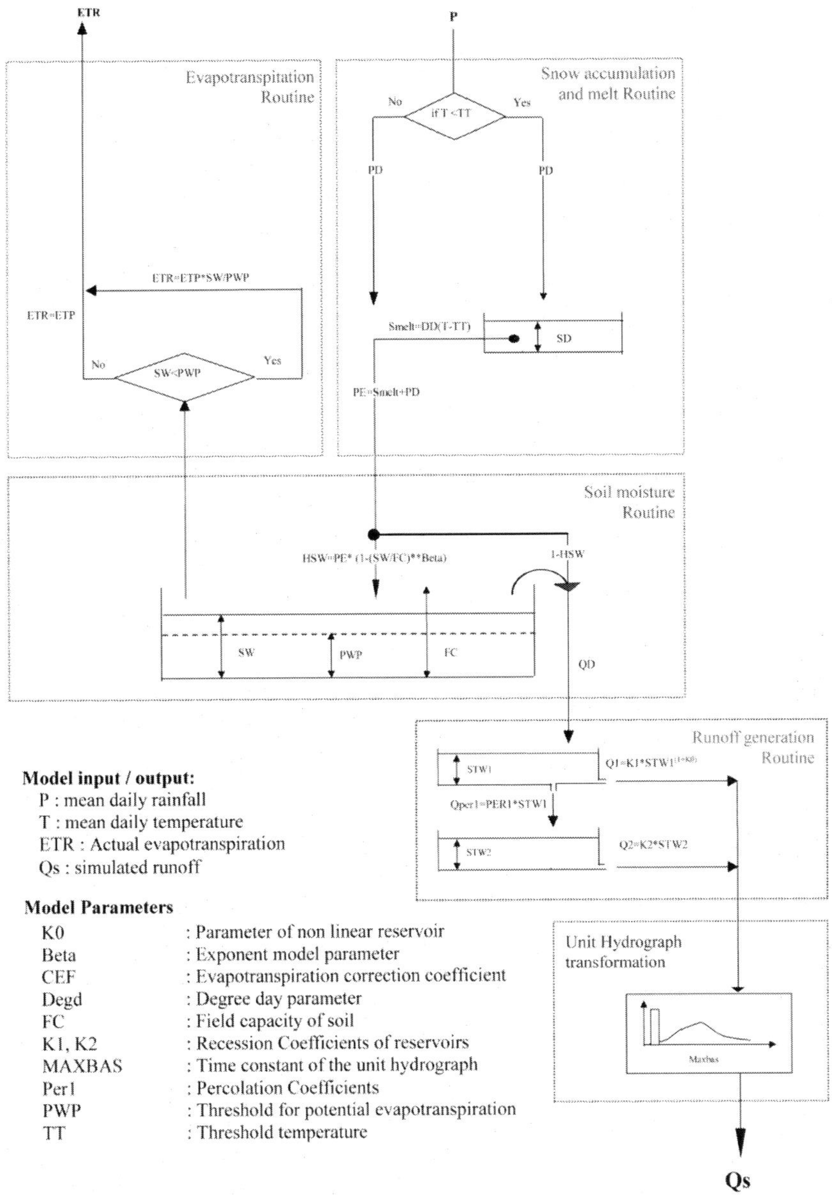

Figure 2. HBV Model structure.

The HBV version we are using is a FORTRAN code of the Institute of Hydraulic Engineering of the University of Stuttgart IWS (Bardossy and Theisen, 1999). Although this version allows a subdivision of catchments under sub-catchments, we adopted a lumped configuration where parameters which are representing the physical and biological properties of the basin (topgraphy, soil and vegetation) as well all some climatic factors are assumed spatially homogenous.

The model time step is hourly, but the results are evaluated at the daily scale, by summarizing outputs over 24 hours. In this application, the model inputs : the daily precipitation, daily temperature and long term average monthly potential evapotranspiration are converted to hourly data. Potential evapotranspiration is the maximum evapo-transpiration rate under given climatic factors, assuming that vegetation is under ideal conditions with respect to soil water content and biological availability. It could be estimated by empirical formula involving air temperture and/or solar radiation. The main outputs are daily mean flows (m^3/s) and daily actual evapotranspiration (mm/day) and daily percolation (mm/day).

The principal components of the HBV model to simulate flux through a soil-atmosphere column are snow accumulation and melt, actual evapotranspiration modeling, infiltration evaluation, soil humidity evolution, and water transfer through soil (Figure 2).

The snow accumulation and melt are first calculated by a degree-day method. Then, the water enters in the soil reservoir where the actual evapotranspiration and percolation are simulated. Two conceptual reservoirs are adopted to transfer percolation amounts through the soil media and to evaluation flux exchanged with the stream. An impulse response function (Beven, 2003) is adopted to simulate the horizontal transfers in order to compute runoff at catchment outlets (unit hydrograph transformation, Figure 2).

In total, 11 parameters are needed to achieve these various computations. They have to be calibrated. The different routines are detailed in the following paragraphs.

2.1.1. The Snow Accumulation and Snowmelt Routine

It applies a degree-day approach. When air temperature is under threshold temperature (TT), considered as the freezing point, the precipitation is considered in the form of snow that accumulates on the soil. Precipitation is considered as rainfall in the other condition. When the temperature increases above TT, snow accumulated will start to melt and contribute to the runoff. The amount of snowmelt is calculated as follows:

$$Smelt(t) = DD*(TT - T(t)) \quad (1)$$

where:

Smelt (t) : snowmelt during the time step t (mm/day)

DD: degree-day factor (mm/°C·day) that reflect the amount of water that melt for every 1°C above the threshold temperature TT.

T (t): average daily temperature in the time step t (°C)

TT : threshold temperature (°C)

2.1.2. The Soil Moisture Routine

This routine controls the soil moisture dynamics. The water balance into the soil is estimated by the following equation:

$$dSW(t)/dt = PE(t) - QD(t) - ETA(t) \quad (2)$$

where:

$Q_D(t)$: runoff formation at the time step t.

SW (t): Soil moisture content

PE (t) : precipitation input (the sum of rainfall and snowmelt)

ETA (t): Actual Evapotranspiration at the time step t, which is calculated by an appropriate routine cited below.

As shown in Figure 2, the soil is represented by a reservoir with a capacity FC (Field Capacity, which is a model parameter related to a hydrodynamic soil property that traduces the maximum moisture that could hold the soil). When this reservoir is filled (SW=FC) the water flux coming from precipitation and snowmelt (PE) goes directly to runoff (Q_D).

When the soil tank is not empty (SW<FC), the proportion of precipitation and snowmelt that pass to runoff (Q_D) is calculated by the following non-linear equation, the remaining go to the soil reservoir.

$$Q_D(t)=\left(\frac{SW(t)}{FC}\right)^{\beta}.PE(t) \tag{3}$$

where

Q_D: runoff formation for the time step t

SW (t): Soil moisture content at the time step t

FC: parameter representing the field capacity

PE: precipitation input (the sum of rainfall and snowmelt)

Eq. 3 controls the sharing of rainfall and snowmelt between the response function (Q_D) which is often called effective precipitation and soil moisture reservoir. The ratio Q_D/PE is also called runoff coefficient.

2.1.3. The Actual Evapotranspiration Routine

The actual evapotranspiration (ETA) includes the evaporation from the basin land surface and lakes and vegetation transpiration to atmosphere. It is calculated on the basis of soil moisture content and the daily potential evapotranspiration. Here a model parameter, PWP representing the permanent wilting point, which is a soil parameter defining the minimum soil moisture content at which the full potential evapotranspiration takes place. At soil moisture contents lower than PWP, it is assumed that actual evapotranspiration is reduced linearly. Moreover, an entirely dry soil moisture condition corresponds to SW=0.

$$\begin{cases} ETA(t) = EPT(t)(SW(t)/PWP) \text{ if } SW(t) < PWP & (4) \\ ETA(t) = EPT(t) \text{ if } SW(t) > PWP & (5) \end{cases}$$

with :

ETA (t): Actual Evapotranspiration at the time step t

SW (t): Soil moisture content at the time step t

PWP: model parameter.

EPT (t): Potential EvapoTranspiration

In the data processing, daily potential evapotranspiration (EPT) is first calculated by correcting the monthly average potential evapotranspiration on the basis of actual mean daily air temperature. Then, the hourly EPT is calculated by dividing the daily EPT by 24.

$$EPT(t)=[1+CEF*(T(t)-AVT)]*PEV \tag{6}$$

where:

AVT: monthly mean long-term air temperature
T: actual daily mean air temperature
CEF : model parameter adopted for correction
AVT: monthly long term mean temperature
PEV : monthly long term average potential evapotranspiration.

2.1.4. The Runoff Generation Routine

This routine transforms the excess of water generated when computing the soil moisture routine into river runoff. This is performed assuming two reservoirs of transfer: the upper, non-linear and the lower, linear. Percolation takes place from the upper reservoir to the lower one. These reservoirs simulate the quick and low flows constituting the two runoff components of the hydrograph.

The upper reconstitutes the quick flow (superficial channels) and the lower reconstitutes the slow flow (base-flow).

The outflows from the upper reservoir consist of direct runoff Q1 and percolation Qper1 (Figure 2):

$$Q1(t) = K1 * STW1(t)^{(1+K0)} \tag{7}$$

$$Qper1(t) = STW1(t) * Per1 \tag{8}$$

where:

Q1 (t) : direct runoff at the time step t
Qper1(t): percolation at the time step t
STW1 (t): water storage in upper reservoir at the time step t
K0: parameter of the upper non-linear reservoir representing the non-linear behaviour
K1: parameter of the upper non-linear reservoir representing a time recession coefficient
Per1: parameter of the upper non-linear reservoir representing a percolation coefficient
The outflow from the lower reservoir consists of the retarded runoff Q2:

$$Q2(t) = K2 * STW2(t) \tag{9}$$

with:

Q2 (t) : retarded runoff at the time step t
K2: parameter of the lower linear reservoir representing a time recession coefficient
STW2 (t) : Water storage in upper reservoir at the time step t.

At each time step, the sum of direct and retarded runoff is transformed by convolution into simulated runoff at the outlet Q_s using the impulse response transformation function (Figure 2).

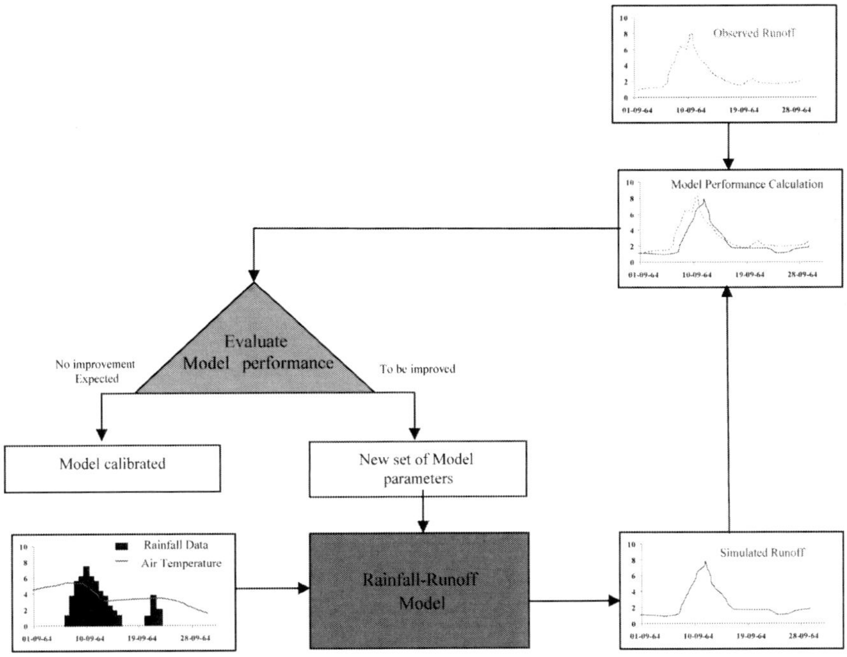

Figure 3. Rainfall Runoff model calibration Schema.

The impulse function is of triangular form including the parameter Maxbas

$$Q_s(t) = \sum_{i=1}^{Maxbas} c(i).Q(t-i-1) \tag{10}$$

where:

$$c(i) = \int_{i-1}^{i} \frac{2}{Maxbas} - \left| u - \frac{Maxbas}{2} \right| \cdot \frac{4}{Maxbas^2} du \tag{11}$$

c(i) : weights of triangular transformation
Maxbas: model parameter
Q_s: discharge at time step t

2.2. Calibration of HBV Model

2.2.1. Statement of Calibration Problem

Now, to be operational for a given watershed, a hydrological model needs to be calibrated on the basis of rainfall and runoff (and/or percolation and evapotranspiration) historical data of this watershed.

A calibration period of generally few years for which input/output information is available is needed. In the present case, the calibration consists on determining suitable model parameter values that gives the best reconstitution of runoff by the model giving the observed rainfall data and air temperature data as inputs (Figure 3).

The calibration of rainfall-runoff model could be considered as an optimization problem stated as follows:

$$\min_{\theta} \{FO\,(Q_{oi}, Q_{si})\,,\ i=1,n\} \tag{12}$$

with:

n : number of time steps of the calibration period.

Q_{oi} : observed runoff at time step i; i=1,... n.

Q_{si} : simulated runoff at time step i; i=1,... n.

θ: model parameters (see table 1 or figure 2 for the case of HBV Rainfall-Runoff model)

FO: objective function.

It minimizes over model parameters (θ), a performance model criterion (FO) that reflects the degree of similarity between observed and simulated runoff series.

For this purpose a choice of a suitable calibration criterion (objective function) and an optimisation algorithm should be achieved.

2.2.2. Calibration Criterion

The selection of relevant model performance criteria to be adopted as objective function remains of primary importance. In the literature, a great number of performance criteria are developed. Generally distance based measures are involved also taking account for model errors heteroscedasticity. In this paper we selected the criterion *RV* (Lindström et al. 1997) as an objective function.

It is a linear combination between Nash-Suttclife criterion (Nash and Sutcliffe, 1970) and absolute relative volume error.

$$RV = \text{Nash} - \omega\,|RD| \tag{13}$$

with:

ω: a weight which is in practice taken equal to 0.1 (Lindström et al. 1997)

RD: relative water balance error term

$$\text{Nash}=1-\frac{\sum_{i=1}^{n}(Q_{si}-Q_{oi})^2}{\sum_{i=1}^{n}(Q_{oi}-\overline{Q}_o)^2} \tag{14}$$

$$RD=\frac{1}{n}\sum_{i=1}^{n}\left(\frac{Q_{si}-Q_{oi}}{Q_{oi}}\right) \tag{15}$$

with

$\overline{Q}_o$: mean observed runoff:

$$\overline{Q}_o = \frac{\sum_{i=1}^{n} Q_{oi}}{n} \tag{16}$$

Nash–Sutcliffe Criteria is a standardised form of squared error (standardised by observed discharge variance). It can range from $-\infty$ to 1. An efficiency of Nash=1 corresponds to a perfect match of modeled runoff to the observed one. The model is more performing when Nash is closer to 1. Popov (1979) considers, on the basis of practical tests, that a model performance is considered "good" for Nash> 0.75, and "satisfactory" for Nash ranging between 0.36 and 0.75.

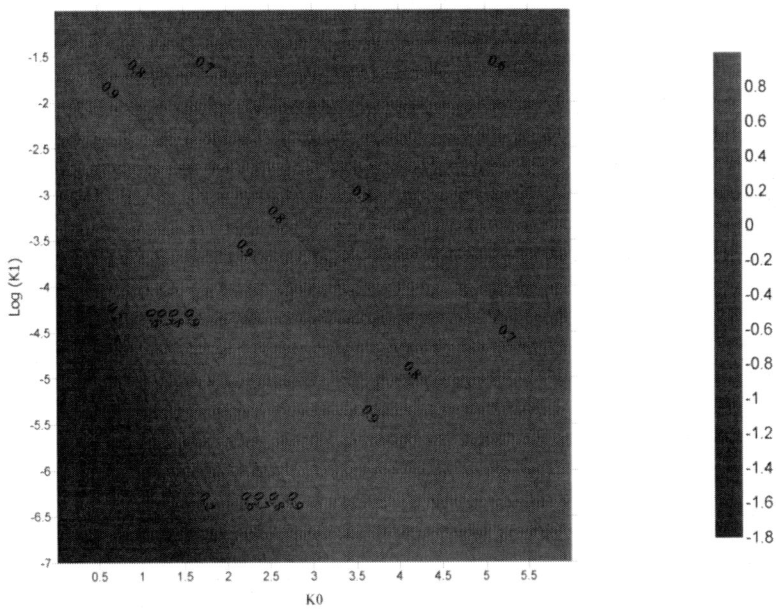

Figure 4. Response surface of the objective function RV in K0- Log(K1) space (the other parameters being fixed).

RD represents the model bias calculated for the whole observation period. RD close to zero indicates that the model does not under-estimate or over-estimate systematically the runoff.

Thus, RV represents a compromise between the Nash-Suttcliffe criterion and relative error. Both the former and the later are of critical importance. In fact the good values of the Nash-Suttcliffe criterion involves a successful reconstitution of hydrographs (short term variability), and a smaller value of RD is obtained for an optimal reconstitution of water balance (long term variability). According to Lindström et al. (1997), RV presents the advantage of procuring good values of Nash with unbiased water balance.

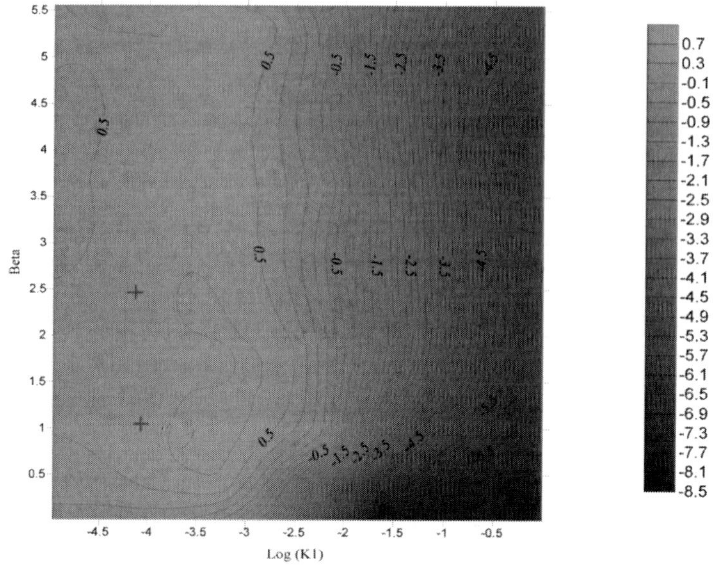

Figure 5. Response surface of the objective function RV in K1-Béta space (the other parameters being fixed), + indicate local optima.

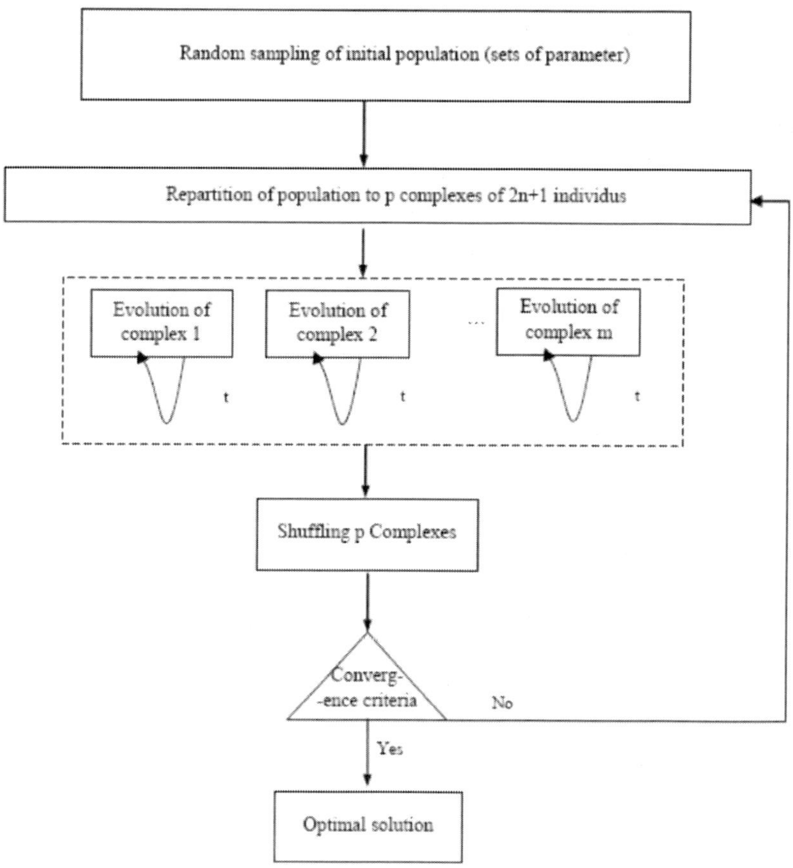

Figure 6. The Shuffled complex evolution algorithm.

2.3. Optimisation Algorithm: SCE-UA

The calibration of Rainfall-Runoff models is recognized to present several difficulties such as interdependence between model parameters as it is reflected from Figure 4 which reports a hill type surface response when considering the pair of parameters K0 and Log (K1).

Other difficulties are the indifference of objective function to some parameters, the difficulty of definition of gradient direction, the existence of local optimums and scale problems related to parameters. As an example, Figure 5 reports local minima highlighted in the surface response in the space of K1- Beta parameters.

The Shuffled Complex Evolution of the University of Arizona (SCE-UA) (Duan et al. 1993) is a global stochastic optimisation method recognized as the most effective optimisation algorithm applied to achieve rainfall runoff model calibration. This optimisation algorithm is adopted in this study to calibrate HBV model.

A general description of SCE-UA optimization method is presented in Figure 6, for the case of minimising objective functions. Typically, the steps towards optimisation are:

1. Random generation of a set of points (individuals) in parameter space, each point representing a value of the parameters vector. Then, the calculation of objective function is performed in every point.
2. Classification of points per increasing objective function.
3. Repartition of individuals into m complexes.
4. Evolution of these complexes in an independent manner, using an evolutionary simplex algorithm.
5. Shuffle of complexes.
6. Verification of convergence criteria. If, satisfactory exit else return to step (2)

According to the literature, SCE-UA takes its robustness, efficiency and effectiveness from the four following concepts on which it's founded: (1) combination between deterministic and probabilistic approaches, (2) systematic evolution of points into a direction which improves the objective function, (3) competitive evolution, and (4) shuffling of complexes.

3. Methodology of SCE-UA Efficiency Improvement Using KNN Implementation

3.1. The K-Nearest Neighbour Technique

The K-Nearest Neighbours (KNN) (Fix and Hodges, 1952; Skellam, 1952) may be considered as a learning technique based on classifying objects according to the closeness to their K neighbours in a feature space, where K is the fixed number of neighbours. The K-Nearest Neighbours estimator is a nonparametric estimator of an n dimensional scalar function F(X), with X a vector of n attributes $(x_1, x_2, .., x_n)$. Given a data set of p realisations of this function, $\{(x_{i1}, x_{i2}, .., x_{in}), F(x_{i1}, x_{i2}, .., x_{in}); i=1,p\}$, this technique gives an estimation of F(X') for X' that could take a different value from sample values. The process is to choose the K-Nearest Neighbours in the feature space $\{(x_{1l}, x_{2l}, .., x_{nl}), l=1,K\}$ where K is the number of neighbour, and to compute a KNN estimation of F(X') as the average of the $F(X_l)$ values.

The KNN method has recently found wide applications in water sciences fields. Khu and Madsen (2005) exploited KNN for the classification of solutions of a multi-objective calibration algorithm using genetic algorithms. Bàrdossy and He (2006) proposed KNN for the regionalization of parameters of the rainfall-runoff model HBV. Ostfeld and Salomons (2005) adopted KNN technique into a genetic algorithm procedure in order to achieve the calibration of an environmental model.

3.2. Method of Implementation of KNN to SCE-UA

Subsequently to a number of iterations generated by SCE-UA, a database of explored sets of variables is constituted. As well, the corresponding objective functions are saved $\{(X_i, FO(X_i)); i=1,p\}$, where p is the number of iterations. Hence, this database is investigated to complete the estimation of the objective function by the K-Nearest Neighbours technique. The application of KNN consists in interpolating the K-Nearest Neighbours using the objective function pattern, for a given new generated point, Xnew.

$$FO_{KNN}(X_{new}) = \frac{1}{\sum_{l=1}^{k} \frac{1}{d_l(X_l, X_{new})}} * \sum_{l=1}^{k} \frac{FO(X_l)}{d_l(X_l, X_{new})} \tag{17}$$

with:

$FO_{KNN}(X_{new})$: objective function estimated by KNN in a new generated point Xnew $(x_{new\,1}, x_{new\,2}, \dots x_{new\,n})$ where x_{newi} (i=1,n) are variable values taken in a point X new

$FO(X_l)$: objective function value at neighbour X_l.

$dl(X_l, Xn_{ew})$: Euclidean normalized distance between neighbour Xl and the new generated point Xnew , with

$$d_l(X_l, X_{new}) = \sqrt{\sum_{i=1}^{n} (\tilde{x}_{newi} - \tilde{x}_{li})^2} \tag{18}$$

with $\tilde{x}_i$: Normalised variable i , defined as follows:

$$\tilde{x}_i = \frac{x_i - x_{i\min}}{x_{i\max} - x_{i\min}} \tag{19}$$

where x_{imax} and x_{imin} are limit values of variable x_i, thus defining the feasible space.

It is obvious that by using the direct simulation model, it provides the required value of the objective function but at the expense of efficiency. Usually, function approximation reduces the computation time but at the expense of accuracy for estimating the objective function. Here we have to make a compromise between the use of approximation function and the direct estimation. In the following, we address the method of choosing the situations where the objective function is estimated by KNN rather than directly.

In Dakhlaoui et al. (2012) it was proposed to choose situations, where objective function evaluation is estimated either by model simulation or by KNN interpolation, using a discrimination of the new generation point taking into account the interdistance to the nearest neighbour in the parameter space. In this study the selection is performed through a discrimination of new generated point considering only the neighbouring objective function values.

Effectively, for every new generated point by SCE-UA the objective function is firstly evaluated by KNN. This point is then ranked with respect to its objective function value in the complex sub-population. Here, the algorithmic parameter p-KNN is introduced as a threshold allowing distinguishing between the "interesting" regions of the feasible space that need to be more finely explored and the "uninteresting" regions, where it is not necessary to spend time for exploration. If this point is one of the p-KNN best points the objective function is calculated by HBV simulation. Otherwise, the approximation by KNN is achieved.

Two additional conditions are assumed. Firstly, if the new generated point is identical to an existent explored point, the objective function value is assigned to the existent. Secondly, if one of the nearest neighbours of the new generated point is the best point or the ranked second best point in the complex or of the whole population, the estimation of the objective function is forced to be by running the model (direct simulation). It's noticed that each time the objective function in a new generated point is estimated by model simulation, this point is stored systematically in the KNN database. Figure 7 reports the different cases of estimation of the objective function by model simulation or by KNN.

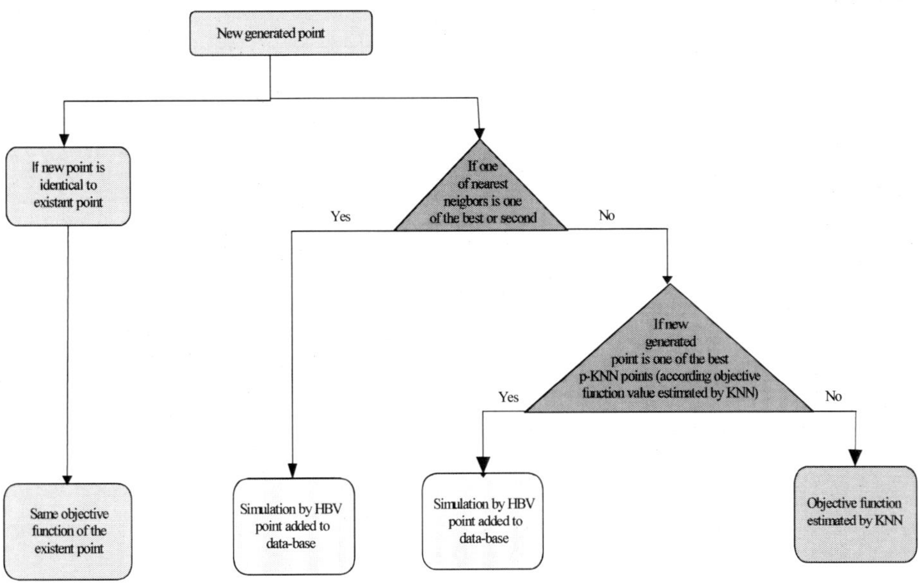

Figure 7. Cases of application of KNN for the estimation of objective function in a new generated point.

4. RESULTS AND DISCUSSION

To quantify the improvement resulting from using KNN for estimating the objective function, synthetic data derived from HBV simulation is performed for the Sejnène basin, a Mediterranean basin situated north Tunisia with a watershed of 378 Km² (Figure 8). Time series of rainfall and daily temperature for the hydrological year 1964-1965 (the period extending from September 1 1964- to August 31 1965) are thus considered as model inputs. The parameter set is obtained by calibration with original SCE-UA. Thus, this set is considered as beforehand optimal solution. Parameters resulting from such calibration are reported in Tab1 together with the assumed parameter range.

The runoff obtained by HBV simulation with this set of parameters is then considered as synthetic input. Rainfall and synthetic runoff series are reported in Figure 9. Rainfall intermittency and high runoff variability are noticeable and constitute one of the difficulties of the calibration process. The synthetic runoff data, free of errors, allows the possibility to deal with a known optimal solution.

Further, the optimisation problem is considered using only six parameters to be calibrated. These parameters are suggested to be more or less sensitive to objective function (Dakhlaoui et al., 2012). The parameter ranges of feasible space derived from the literature review are shown in Table 1.

Ten independent runs of the initial SCE-UA (Duan et al. 1993) and the enhanced one (SCE-UA with KNN implementation) are performed. For each run the initial population is generated randomly. The number of Complexes in SCE-UA is fixed to 4, the population per complex is chosen equal to 13, as recommended by Duan et al. (1994). Preliminary tests have shown that the suitable number of Neighbour to be adopted for KNN technique is K=3.

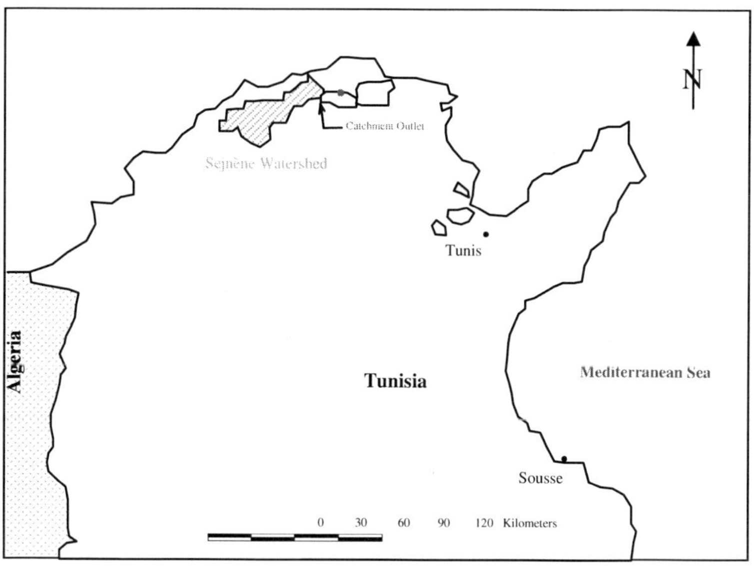

Figure 8. Geographical location of Sejnène catchment.

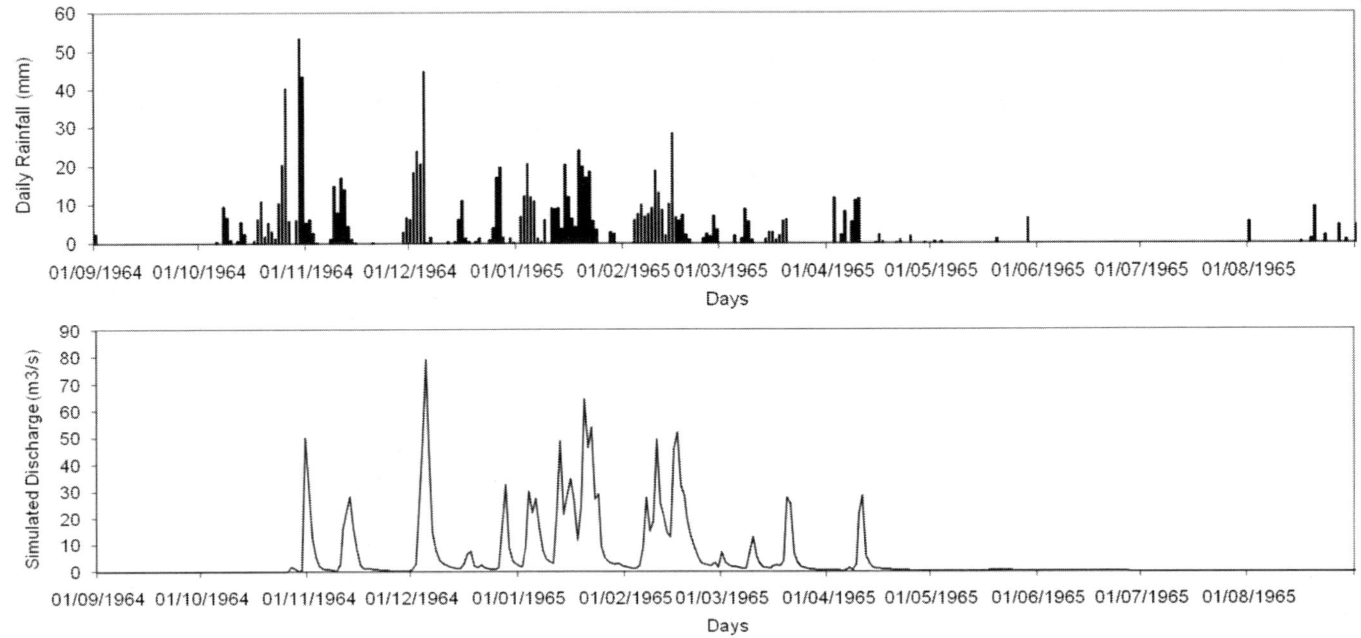

Figure 9. Synthetic data of Sejnène for the period September 1964-August 1965 : observed rainfall and simulated discharge by HBV model within parameter set (Table 1).

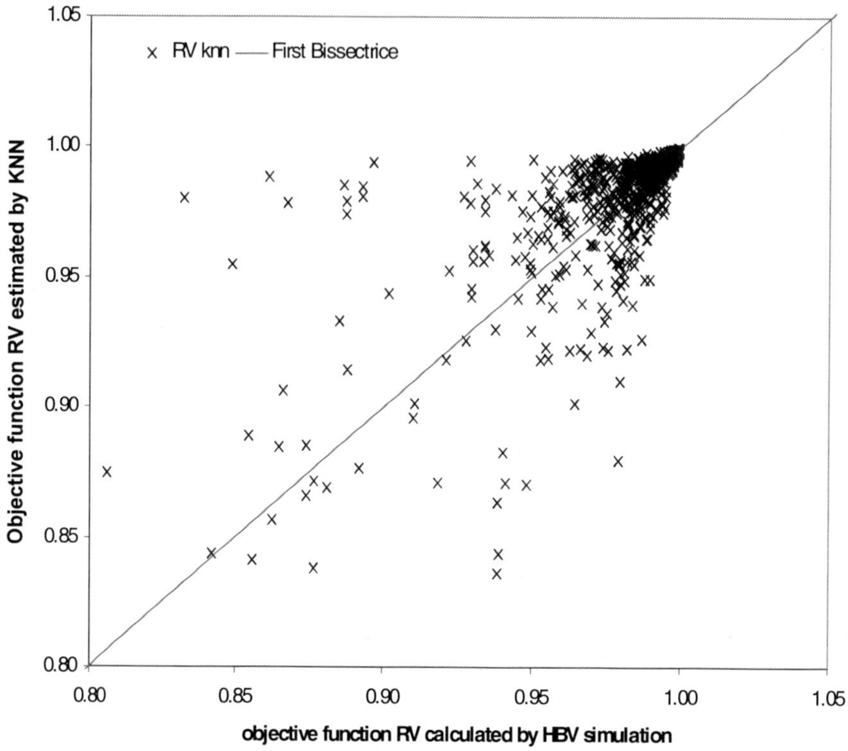

Figure 10. Objective function estimated by KNN compared to objective function calculated by model simulation.

From Figure 10, it may be concluded that the KNN estimation is closer to the objective function estimated by model simulation, for large values of *RV*. This may suggest that the KNN involvements are more accurate for the most interesting zones of the research space. Hence, This KNN approach helps to preserve the accuracy of the optimization algorithm near the optimal solution.

In Figure 11a the optimal parameters are reported for the case of original SCE-UA, using ten independent runs (replicates) of SCE-UA. For comparison purposes the values are linearly normalized (so that the width of the parameter range equal to 1 and the synthetic optimal solution equal to 0.5 which is reported by a red scattered line). In Figure 11b to Figure 11g, we study the effect of the modification of the algorithmic variable p-KNN. Five values are considered from 5 to 50. For each value, ten independent runs of modified SCE-UA are also performed and each time, the optimal set is identified.

Optimal parameters obtained by Modified SCE-UA are reported in Figure 11b, c, d, e and f, run by run, and compared to those deriving from the initial algorithm reported in Figure 11a.

It may be seen that where p-KNN is greater than 15, the effectiveness of the algorithm begins to be deteriorated giving the fact that the algorithm doesn't converge toward the true solution.

Table 1. Feasible space of HBV model and synthetic optimal solution for Sejnène Only Bold parameter are calibrated

Parameter	Min	Max	Optimal Synthetic Solution
K0	0	3	2.93
Log (K1)	-7	0	-5.64
Parameter	Min	Max	Optimal Synthetic Solution
Log (K2)	-7	0	-6.54
Log (PER1)	-7	-1	-2.75
TT (°C)	-1.5	2.5	-
DEGD (mm/°C)	1	0.0	-
MAXBAS (h)	1	48	13
PWP (mm)	10	500	61
FC (mm)	10	500	282
BETA	0	6	1.62
CEF	0	0.1	0.013

This value of p-KNN = 15 (case 4 , Figure 11b) will be finally retained in the modified algorithm. Each time, the number of iterations to converge by modified SCE-UA is computed. The percent of reduction of the number of iterations required to convergence when the p-KNN changed is then evaluated and reported in Figure 11g. The values of objective function reached at the optimal solution for the different cases of p-KNN values seems to be close to 1 (Figure 11h). We note that increasing p-KNN, the number of iterations of SCE-UA to reach the optimal solution is reduced which means that efficiency is improved. Adopting p-KNN=15, we may draw the conclusion that KNN implementation gives an improvement in efficiency of the optimisation algorithm of about 38%, compared to initial algorithm, at the same time conserving an identical rate of effectiveness.

Table 2. Differents Cases of p-KNN values

Cases	p-KNN
1	Without KNN
2	10
3	20
4	30
5	40
6	90

Although, with p-KNN greater than 15, the sample distributions of parameter values taken at the optimal solution are found quite different from the true solution, the enhanced algorithm converges toward solutions with RV close to the true solution with uncertainty of about 2 ‰ (RV between 1 and 0.998).

This accuracy in objective function could be considered as acceptable for engineering purposes. Thus, it's possible to obtain an enhanced calibrating algorithm 2 to 3 times faster than the initial SCE-UA.

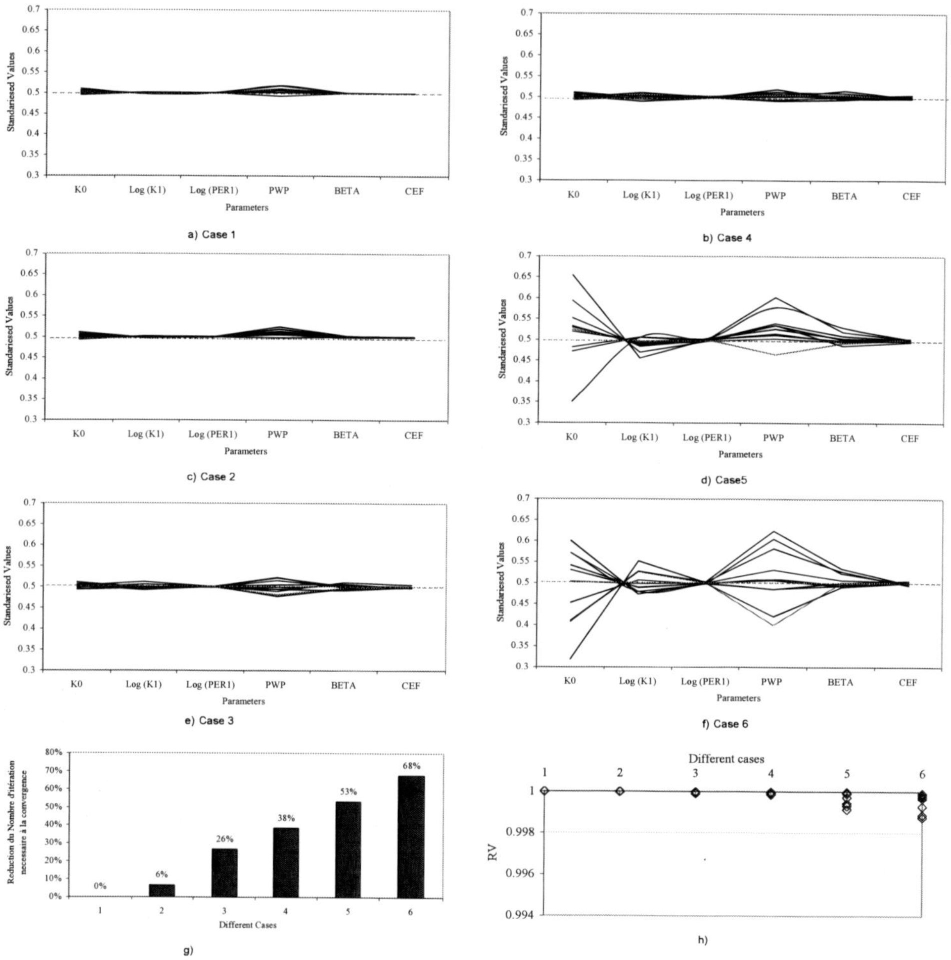

Figure 11. Effect of the modification of p-KNN on the number of iterations to convergence and values of the objective function and parameters taken at optimum. Every line joins the normalized values of parameter at the optimum solution of a run of SCE-UA. Scattered line is the synthetic solution. p-KNN values for each case are reported in Table 2.

CONCLUSION

In this study we have tried to continue our previous works on improving the efficiency of SCE-UA by using the KNN technique as an alternative estimator for the objective function during the optimization process. The implementation of KNN to help estimate the objective function in specific optimisation stages allows the possibility to SCE-UA to make use of all the information about the explored points by prior model iterations. The adoption of the objective function computation by KNN interpolation from previous estimations was tested in the present. However, to reduce the effect of KNN estimation error, the assessment of objective function by model simulation was restricted according a discrimination of new generated point among a ranking of the explored population by their objective function values.

On the basis of synthetic discharge data adopted using HBV rainfall-runoff model with only six unknown parameters, the conclusion can be drawn that the efficiency of the optimisation algorithm could be improved by about 40 % compared to the initial one. In our previous work, where the selection of situations where objective function is estimated by the model simulation rather than by KNN was achieved through a discrimination using threshold distances between the newly generated individual and its neighbours, the improvement was 30% (Dakhlaoui et al., 2012).

More attention is intended to be assigned in future works to the definition of the recommended values of the algorithmic KNN parameters (p-KNN and K). Another perspective could be the investigation of other distance measures in KNN. Also, it would be important to study the effect of the choice of the parameters involved in the optimization process. Here, the parameter FC representing the soil field capacity was not part from the calibration. In fact, the parameters interdependency may play a key role in calibration performances.

REFERENCES

Bates B.C., 1994. Calibration of the SFB model using a simulated annealing approach. *Proc. Int. Hydrol. and Water Resour. Sym*p., Adelaide, Vol.3, 1-6.

Begström S., 1976. Development and application of a conceptual rainfall-runoff model for the Scandinavian catchments. *SMHI RH07*, Norrköping.

Bardossy A. and Theisen H.W., 1999. Listing en Fortran du modèle HBV, Version 0.3. IWS, Stuttgart.

Beven K.J., 2003. Rainfall – Runoff modelling: The Primer. Wiley. ISBN-10: 0470866713 | ISBN-13: 978-0470866719.

Dakhlaoui H., Bargaoui Z., and Bardossy A. 2012. Toward a more efficient Calibration Schema for HBV Rainfall-Runoff Model. Journal of Hydrology 444-445 (2012) 161-179. http://dx.doi.org/10.1016/j.jhydrol.2012.04.015.

Duan Q., Gupta V. K., and Sorooshian S. ,1993. A Shuffled complex evolution approach for effective and efficient global minimization. Journal of optimisation *Theory Application,* 76 (3) , 501-521.

Duan Q., Sorooshian, S. and Gupta, V., ,1994. Optimal Use of the SCE-UA Global Optimisation Method for Calibrating Watershed Models. *Journal of Hydrology*, No. 158, 265-284.

Efstratiadis A. and Koutsoyiannis D., 2002. An evolutionary annealing-simplex algorithm for global optimisation of water resource systems. Hydroinformatic 2002: *Proceedings of the fifth international conference on hydroinformatics,* Cardiff UK. IWA Publishing , pages 1423-1428.

Fix E. and Hodges J.L., 1952. Discriminatory Analysis: Non parametric Discrimination: Small Sample Performance. project 21 - 49 – 004, Report Number 11 USAF School of Aviation Medicine, Randolph Field, Texas, 28 – 32.

Franchini M., 1996. Use of genetic algorithm combined with a local search method for the automatic calibration of conceptual rainfall-runoff models. *Hydrol. Sci. J.*, 42 (3), 357 - 379.

Franchini M., Galeati G., and Berra S., 1998. Global techniques for the calibration of conceptual rainfall-runoff models. *Hydrological Science Journal* 43, 443-458.

Goldberg D.E. 1989. Genetic algorithms in search, optimization and machine learning. Reading, MA: Addison-Wesley.

Giannakoglou K. C., 2002. Design of optimal aerodynamic shapes using stochastic optimisation methods and computational intelligence, *International Review Journal Progress in Aerospace Sciences*, vol. 38, no. 5, pp. 43–76.

Gupta H.V., Sorooshian S., and Yapo P.O., 1999. Status of automatic calibration for hydrologic models: comparison with multilevel expert calibration. *Journal of hydrologic Engineering,* April 1999 , 135-143.

Khu T. S., and Madsen H., 2005. Multiobjective calibration with Pareto preference ordering: An application to rainfall-runoff model calibration, *Water Resour*. Res. 41.

Kuczera G., 1997. Efficient subspace probabilistic parameter optimization for catchment models. *Water Resour. Res.*, 33, 177–185.

Lindström G., Johanson B., Gardelin M.P.M., Bergström S., 1997. Developpement and test of the distributed HBV-96 hydrological model. *Journal of hydrology* 201 (1997) 272-288..

Nash, J. E., and Sutcliffe J. V., 1970. River flow forecasting through conceptual models. 1. A discussion of principles. *J. Hydrol.,* 10 (3), 282–290.

Ong Y. S., Nair P. B., and Keane A. J., 2003. Evolutionary optimization of computationally expensive problems via surrogate modelling. *American Institute of Aeronautics and Astronautics Journal*, vol. 41, no. 4, pp. 687–696.

Ostfeld A. and Salomons S., 2005. A hybrid genetic-instance learning algorithm for CE*QAL-W2 calibration. *Journal of Hydrology* 310 (2005) 122-125.

Popov, E.G. *Gidrologicheskie Prognozy (Hydrological Forecasts).* Gidrometeoizdat.: Leningrad, Russia, 1979.

Skellam J.G., 1952. Studies in statistical ecology. I. Spatial pattern, Biometrica,39,346-362.

Tolson B.A., and Shoemaker C.A., 2007. Dynamically dimensioned search algorithm for computationally efficient watershed model calibration. Water Resources Research, vol. 43, W01413, doi:10.1029/2005WR004723.

Thyer M., Kuczera G., and Bates B. C., 1999. Probabilistic optimisation for conceptual rainfall-run-off models: A comparison of the shuffled complex evolution and simulated annealing algorithms. *Water Resources* Research 35(3),767-773.

Wang Q.J., 1991. The genetic algorithm and its application to calibrating conceptual rainfall-runoff models. *Water Resources Research*, 27(9), 2467-2471.

Zhou Z., Ong, Y. S. Nair, P. B., Keane, A. J., and Lum, K. Y., 2007. Combining global and local surrogate models to accelerate evolutionary optimization, IEEE Transactions on Systems, Man and Cybernetics (SMC), part C - Applications and Reviews, vol. 37, no. 1, Jan, 2007.

In: Calibration Technology, Theories and Applications ISBN: 978-1-62618-808-2
Editors: Ikumatsu Fujimoto and Kunitoshi Nishimura

Chapter 6

ADVANCEMENT IN CALIBRATION AND VALIDATION OF A MODEL USING DATA DEPTH FUNCTION

Shailesh Kumar Singh*
National Institute of Water and Atmospheric Research, Riccarton
Christchurch, New Zealand

Abstract

A model is simplified mathematical representation of very complex processes. Any model needs parameters to represent the mathematical relationship and the physical processes. Sometimes these parameters can be directly determined from the physical processes, but it is not always the case. For example, only some of the parameters of a physically based hydrological model can be derived from the catchment characteristics. Hence, we need parameters to be estimated before the model being used. The process of training the model based on observation is called calibration. In recent decade due to increase in computational power there has been tremendous advancement in parameter estimation techniques. There are two types of model calibrations, namely manual calibration and automatic calibration. In former calibration the parameters are adjusted by visual inspection and the success of this method depends on the expertise of the modeler. It is also a time consuming and tedious job. Hence, nowadays automatic calibration is preferred. In automatic calibration some kind of optimization algorithm is used to maximise or minimise an objective function. But non-uniqueness of the parameters makes parametrisation a challenging task. There are several local and global optimisation algorithms available. Some of the algorithm results depend on initial guesses and can be trapped in local minima or maxima. So, to overcome such types of problems, global optimisation algorithms like Shuffled Complex Evolution-University of Arizona (SCE-UA), Simulated Annealing (SA), Genetic algorithm (GA), Dynamically Dimensioned Search (DDS), Robust Parameter Estimation (ROPE) algorithm etc. have emerged. A very recent technique like ROPE algorithm uses data depth function to reach optimal parameter space instead of parameters set. Data depth function is a kind of multivariate ranking. It has not been used much in hydrology or water

*E-mail address: shailesh.singh@niwa.co.nz

resources problems. But data depth function is a powerful tool due to its robust properties. It can be used for different purposes in calibration and validation of hydrological models.

In this chapter it has been described how tools like data depth function can advance the calibration of hydrological models. This chapter provides a brief review of the problems associated with parameter estimation of the hydrological model and its possible solutions using data depth function.

Keywords: Hydrological model, parameters, calibration, validation, data depth function, ROPE algorithm, TopNet

1. Introduction

A model is simplified mathematical representation of very complex processes. A model can be very simple to very complex, based on the actual process it represents. For example a hydrological model is a mathematical tool to represent a very complex and nonlinear natural phenomenon (hydrological processes in hydrological cycle) in simple ways. The hydrological model has been in use for several decades for several purposes (flood forecasting, watershed management etc). Even so, hydrological science is not complete in itself because we have limitation in understanding some complex and nonlinear natural processes. Several decades of research has been devoted to make a simple hydrological model for every case study, but with limited success.

Even though the twentieth century has witnessed an enormous growth in hydrological process understanding, numerous challenges still remain (Jakeman and Hornberger, 1993; Blöschl and Sivapalan, 1995; Bergström and Graham, 1998; Singh and Woolhiser, 2002; Moradkhani et al., 2005; Bárdossy and Singh, 2008; Singh, 2010; Singh et al., 2012a). Among the challenges that remain, the dominating ones are: (1) Representation of dynamic hydrological process by static model; (2) Complexity in hydrological model; (3) Accuracy in measurement; (4) Gap between model and real world; (5) Scale problem; (6) Parameterization of model; (7) Hydrological process in the soil micro pores; (8) Preferential flow at large scale, and many more.

The study of hydrological models are two-fold: models for understanding physical processes and models for prediction. In the former, we try to represent the very complex and nonlinear process by some mathematical equation and try to learn at the small scale. In the latter, modelers attempt to predict the stream flow at some future time when given knowledge of the current state of the system and model parameters. In this respect, good estimates of the parameters and state variables are needed to enable the model to generate accurate forecasts (Moradkhani et al., 2005).

Broadly there are three types of hydrological models: empirical, conceptual and physically based (Wagener et al., 2004). At this advanced stage of hydrology, irrespective of the empirical, conceptual or physical type, a hydrological model cannot describe the precise nature of rainfall-runoff processes. This is mainly due to the lack of very clear and sound understanding of physics behind rainfall-runoff process and in adequate representation of the available process knowledge (Patil, 2008). A great deal of research has taken place in the past decades to improve our basic understanding of hydrological processes and implementing the knowledge in hydrological modeling, which in turn has resulted in increased model complexity (Blasone, 2007). The vast complex nature cannot be represented by present knowledge. The current state of knowledge does not allows for complex interaction among various hydrological variables and presently employed formulations to be characterized in the model structure (Sivapalan et al., 2003). A common generalization is that the complex model can represent hydrological processes in a better way, but several authors (Michaud and Sorooshian, 1994; Refsgaard and Knudsen, 1996; Perrin et al., 2001, 2003; van der Linden and Woo, 2003) have proved that the complex hydrological model need not be the best, a simple model can also represent better. However, there is no fixed framework for building a model that is simple in a certain sense. Although, we cannot understand the very complex nature of rainfall-runoff process precisely, hydrological models are still of great use for any water resources developmental activity. Any model before its real use needs to be calibrated to obtain its parameters. This makes parameterization an important problem. Note that not only the hydrological model suffer from the above mention problems but any kind model suffers from similar problems.

All the hydrological models have a certain number of parameters; some of them have physical meaning, while others do not. The only problem with hydrological models are, that they need to estimate their parameters using observed data before the model can be used for practical purposes. This is because the predictability of the model will very much depend on parameters. The typical way to estimate the parameters is trying and adjusting the parameters values by various means so that precipitation-runoff behavior of the model approximate as closely and consistently as possible to the observed response of the catchments over some period of time for which precipitation, stream flow and other relevant catchments characteristic measured data are available. This process of parameter estimation is called model calibration (Gupta et al., 2003). Several techniques have been developed in different corners of the world to estimate parameters of a hydrological model. Generally, there are two types of model calibration. One constitutes the manual calibration, which completely relies on the expertise and judgment of the hydrologist. The other method is automatic calibration, which employs the power, ability to follow systematic programed rules, speed and capability of the computer (Gupta et al., 2003; Hendrickson et al., 1988).

In manual calibration, the evaluation of the agreement between simulated and observed hydrograph is subjective and based on visual inspection. The parameters are tuned based on the expert guesses. For the automatic calibration, a different single or multi-objective function for the parameter adjustments and different criteria can be used for evaluation of goodness of fit between observed and modeled hydrograph. The non-uniqueness of the parameters makes parametrization a challenging task. There are several local and global optimization algorithms available. Some of the algorithmt's results depend on the initial guesses and can be trapped in local minima or maxima. So to overcome such type of problem, global optimization algorithms like SCE-UA, Simulated annealing, Genetic algorithm etc. have emerged (Kavetski et al., 2006b). The advantages and disadvantage of manual and automatic calibration are given in Table 1 as given by Gupta et al. (2003); Singh (2010).

There are several problems associated with parameter estimation. The major problems were described by Jackson and Aron (1971); Singh (2010):

- Which method is best suited for analysis?
- Criteria used for

Table 1. The advantages and disadvantages of manual and automatic calibration

Manual Calibration	Automatic Calibration
User knowledge and expertise is valuable	Speed and computational power of computer
Very subjective	Based on objective function
Labor intensive	Computational intensive
Time consuming	Time saving
Excellent result	Good result; some time result not acceptable
Not suitable for long time period of calibration	Can calibrate using long time period

evaluating the parameters

comparison of model result with observed data (goodness of fit)

- Several kind of errors associated with observed data
- Problem associated with structure of the model
- Integration and combination of the above problems
- Uncertainty associated at various level of parameter estimation
- Availability of data
- Initial and boundary condition

Irrespective of this, all model calibrations and their subsequent predictions will be subjected to uncertainty (Beven, 2000) due to above problem associated with parameter estimation. Parameter estimation for hydrological models has received increased attention from the hydrology and land surface modeling community (Sorooshian et al., 1993; Kuczera, 1997; Gupta et al., 1998; Andréassian et al., 2001; Beven and Freer, 2001; Vrugt et al., 2002; Gupta et al., 2003; Wagener et al., 2003; Samaniego and Bárdossy, 2005; Beven., 2006; Kavetski et al., 2006a,b; Bárdossy, 2007; Bárdossy and Singh, 2008). In the process of model calibration, we have to obtain our model parameter from feasible space of parameter using some kind of objective function. But it is often very complex and difficult to find the response surface in the parameter space (Beven, 2000; Beven and Freer, 2001). Certain extend we can overcome these difficulties with a simple model with small number of parameters. This can have a smoother parameter response surface, but in reality it is very difficult to archive in hydrological modeling (Beven, 2000).

In this chapter, we demonstrate how concepts of geometry and multivariate statistics are used to address the problem of parameter estimation or model calibration.

2. Calibration and Validation

Calibration of a hydrological model can be defined as the process of adjusting or tuning the parameters of a hydrological model such that model output is approximate as closely and consistently as possible to the observed output value and represent reasonably the hydrological processes of the watershed. The process of calibration is systematically represented in Figure 1. During calibration we need model inputs, model parameters, observed values, an objective function, and an optimization algorithm. The objective function is a mathematical expression to compare quantitatively the observed value with model output. The optimization algorithm is one which generates parameters based on objective function values. Once the objective is reached, corresponding parameters set is called calibrated parameters. After the calibration, it is very important to test our model on another set of data where model was not calibrated. So testing of any calibrated model on completely new data where model was not calibrated is called validation. A systematic representation of the process of validation is given in Figure 2. The model is ready to use if it works well on validation data.

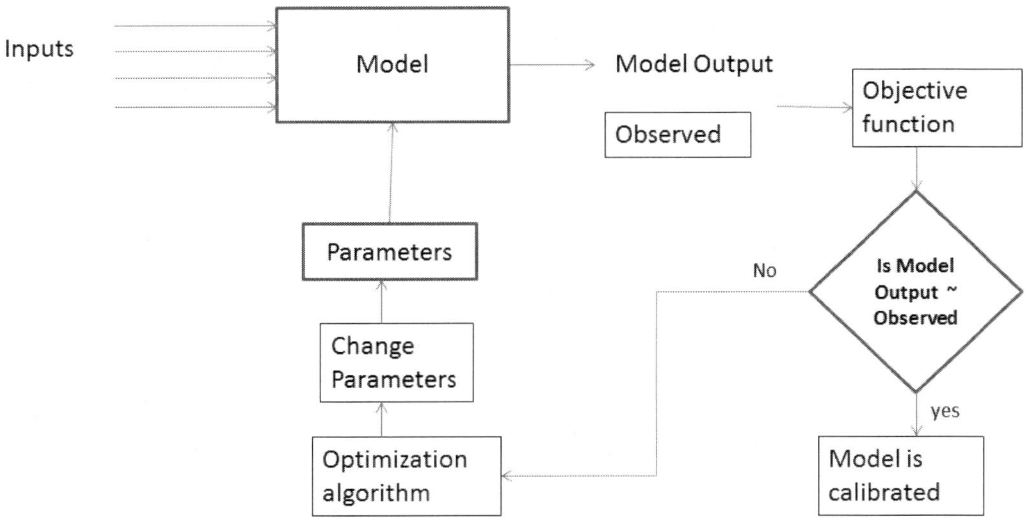

Figure 1. Systematical representation of the process of calibration.

In the following section a calibration and validation of a hydrological model with details about the objective function and optimization algorithm is given.

2.1. Objective Functions

There are several objective functions available for evaluating the closeness of model output with observed value. The most widely used objective functions are Nash-Sutcliffe coefficient (Nash and Sutcliffe, 1970), Root mean square, Volume error and Peak error etc. The equation 1 to 4 describes the formulation of above mentioned objective functions. Generally extreme value can dominate in the above mentioned objective functions, hence, to normalized the effect of the extreme value, log of the above mention objective can be defined. To calculate log objective function, model output and observed values are replaced by log model output and observed value.

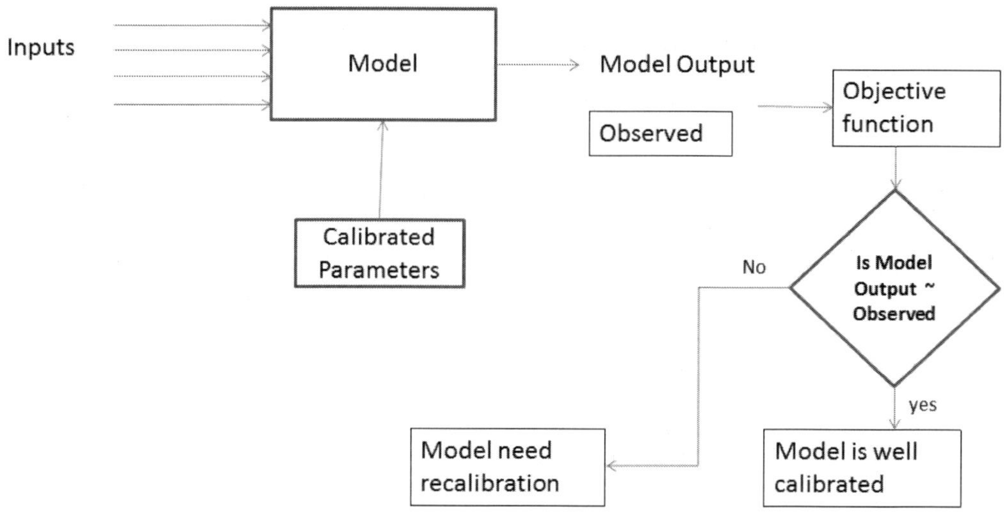

Figure 2. Systematical representation of the process of validation.

$$NS = 1 - \sum \frac{(X_o - Y_m)^2}{(X_o - \widehat{X}_o)^2} \tag{1}$$

$$RMSE = \left[\frac{\sum (X_o - Y_m)^2}{N} \right]^{1/2} \tag{2}$$

$$VE = \frac{1}{N} \sum \left| \frac{(Y_m - X_o)}{X_o} \right| \tag{3}$$

$$PE = \frac{1}{N} \sum \left| \frac{(Y_{m(peak)} - X_{o(peak)})}{X_{o(peak)}} \right| \tag{4}$$

Where NS is Nash-Sutcliffe coefficient, X_o is the observed value, Y_m is model output. RMSE is root mean square, N is the simulation length. VE is volume error, PE is peak error, $Y_{m(peak)}$ is peak model output, $X_{o(peak)}$ is observed peak. Peak values can be defined based on threshold value defined by modeller.

2.2. Optimization Algorithm

There are several optimization algorithms available. To explain the concept of calibration of a hydrological model, in this chapter Robust Parameter Estimation (ROPE) algorithm which uses the notion of data depth function is used.

2.2.1. Robust Parameter Estimation (ROPE) Algorithm

ROPE algorithm (Bárdossy and Singh, 2008) is a new methodology for model calibration. ROPE algorithm looks for optimal parameter space instead of a single parameter set.

It uses the merit of data depth function defined by Tukey (1975). Data depth is a quantitative measurement of how central a point is with respect to a multivariate data cloud or a distribution. This gives us the central outward ordering of a data points in a multivariate sample. This center-outward ordering of the multivariate data provides a foundation for various multivariate estimation and inference methods. Depth functions were first introduced by Tukey (1975) to identify the center (a kind of generalized median) of a multivariate dataset. Several generalizations of this concept have been defined in Rousseeuw and Struyf (1998), Liu et al. (1999) and Zuo and Serfling (2000), Liu (1990), Donoho and Gasko (1992). The Tukey's half-space depth of a point p with respect to the finite set X in the d dimensional space $\Re^d$ is defined as the minimum number of points of the set X lying on one side of a hyperplane through the point p. The minimum is calculated over all possible hyperplanes. Half-space depth function gives us depth of a points with respect to a space or a distribution. The depth tell us how central is the points with respect to the given space or distribution.

Formally the half-space depth of the point p with respect to set X is define by equation 5 (Bárdossy and Singh, 2008).

$$D_X(p) = \min_{n_h \in X \subset R^d} \left(\min_{x \in X} \left(|x| : x \in X \langle n_h, x-p \rangle > 0 \right), \left(|x| : x \in X \langle n_h, x-p \rangle < 0 \right) \right) \quad (5)$$

Here $\langle x, y \rangle$ is the scalar product of the d dimensional vectors, and n_h is an arbitrary unit vector in the d dimensional space representing the normal vector of a selected hyperplane. If the point p is outside the convex hull of X then its depth is 0. Points on and near the boundary have low depth while points *deeply* inside have high depth. In higher dimensions, the calculations of depth functions is quite computationally intensive. The points with high depth are the points which lie in the interior of the data cloud while those with low depth lie near the side of the convex hull. In one dimension data depth will be nothing but the rank. The convex hull of a set of points S is the smallest area polygon which encloses S. Mathematically, The convex hull of a set of points S in d dimensions is the intersection of all convex sets containing S. Let take N points $p_1, p_2, ...P_N$, then the convex hull S can be define by equation 6 (WolframMathWorld, 2012).

$$conv(S) = \left\{ \sum_{i=1}^{N} \alpha_i p_i | \forall i : \alpha_i \geq 0, and \sum_{i=1}^{N} \alpha_i = 1 \right\} \quad (6)$$

Convex hull S is the set of all convex combinations of its points. Each point in a convex combination,is assigned a coefficient in such a way that the coefficients are all positive and sum to one. The coefficient are used to compute a coefficient average of the points. For each choice of coefficients, the resulting convex combination is a point in the convex hull, and the whole convex hull can be formed by choosing coefficients in all possible ways (Wikipedia, 2012).

An example of a convex hull in two dimensional space in practical case is given in Figure 3. The X axis of the Figure 3 is one parameter of a model and Y axis is another parameter. All the points that lie inside the convex hull have higher depth.

There are certain properties of data depth functions which makes them very powerful tools in providing a center-outward ordering of points in a multivariate dataset. It should satisfy following basic properties (Zuo and Serfling, 2000):

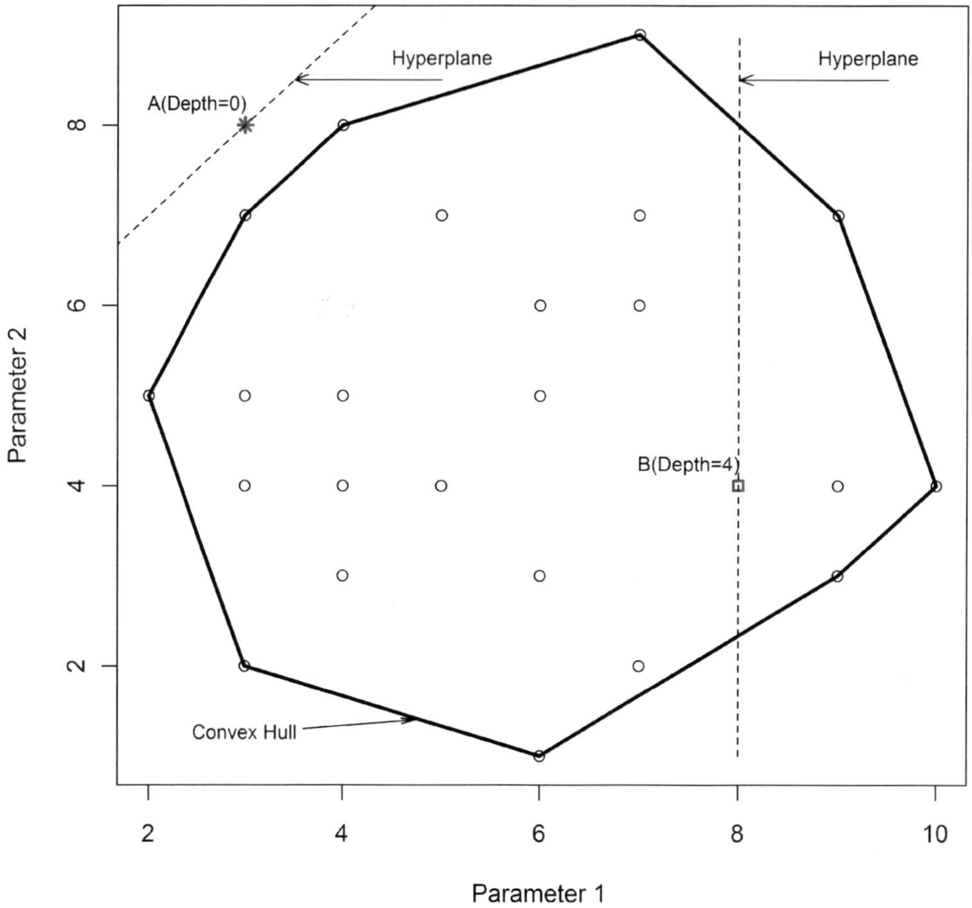

Figure 3. Example of convex hull.

- *Affine invariance*: The depth of a point $X \in \Re^d$ should not depend on the underlying coordinate system or, in particular, on the scales of the underlying measurements.
- *Maximality at center*: For a distribution having a uniquely defined center (e.g., the point of symmetry with respect to some notion of symmetry), the depth function should attain maximum value at this center.
- *Monotonicity relative to deepest point*: As a point $X \in \Re^d$ moves away from the deepest point (the point at which the depth function attains maximum value; in particular, for a symmetric distribution, the center) along any fixed ray through the center, the depth at X should decrease monotonically.
- *Vanishing at infinity*: The depth of a point X should approach zero as X approaches infinity.

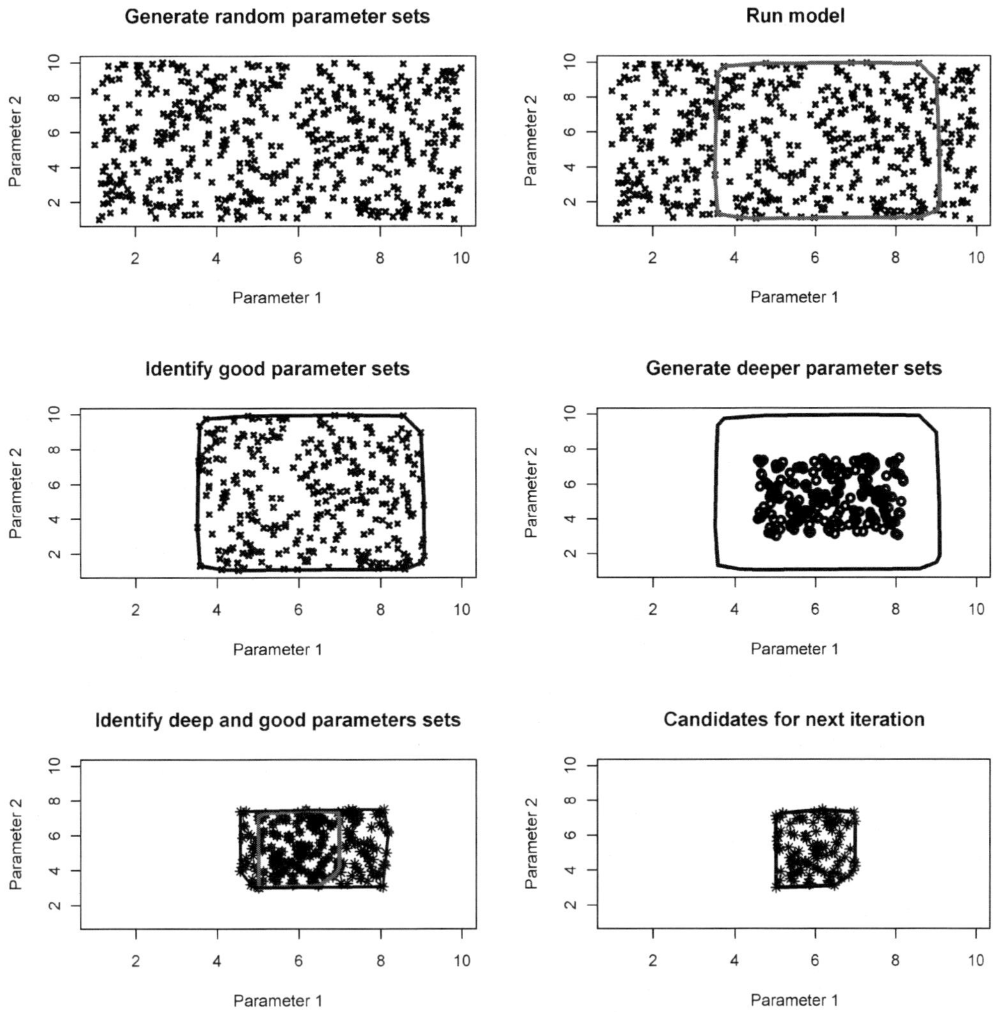

Figure 4. Systematic representation of ROPE algorithm (Singh, 2010).

ROPE algorithm uses these robust properties of data depth function to estimate an optimal parameter set. Brief of the ROPE algorithm is given below; a detailed treatment of the ROPE algorithm can be found in Bárdossy and Singh (2008). Figure 4 contains a systematic representation of the ROPE algorithm. First, a large number of parameter sets are generated from uniform distribution in the feasible parameter space as shown in Figure 4(a). Then the model is run for all the parameters and the best **N** percent of parameter sets are selected as shown in Figure 4(b). After removing the parameters outside the convex hull as shown in Figure 4(c), another set of same number of parameters is generated such that it has higher depth and the parameters are within the convex hull as shown in Figure 4(d). Again the model is run and the performance is calculated and the best **N** percent parameters are selected. The cycle of iteration is continued till the pre-determined number of iterations is over or the variation in performance is within a selected range. For more details about the algorithm please refer to (Bárdossy and Singh, 2008; Singh, 2010; Singh et al., 2012b).

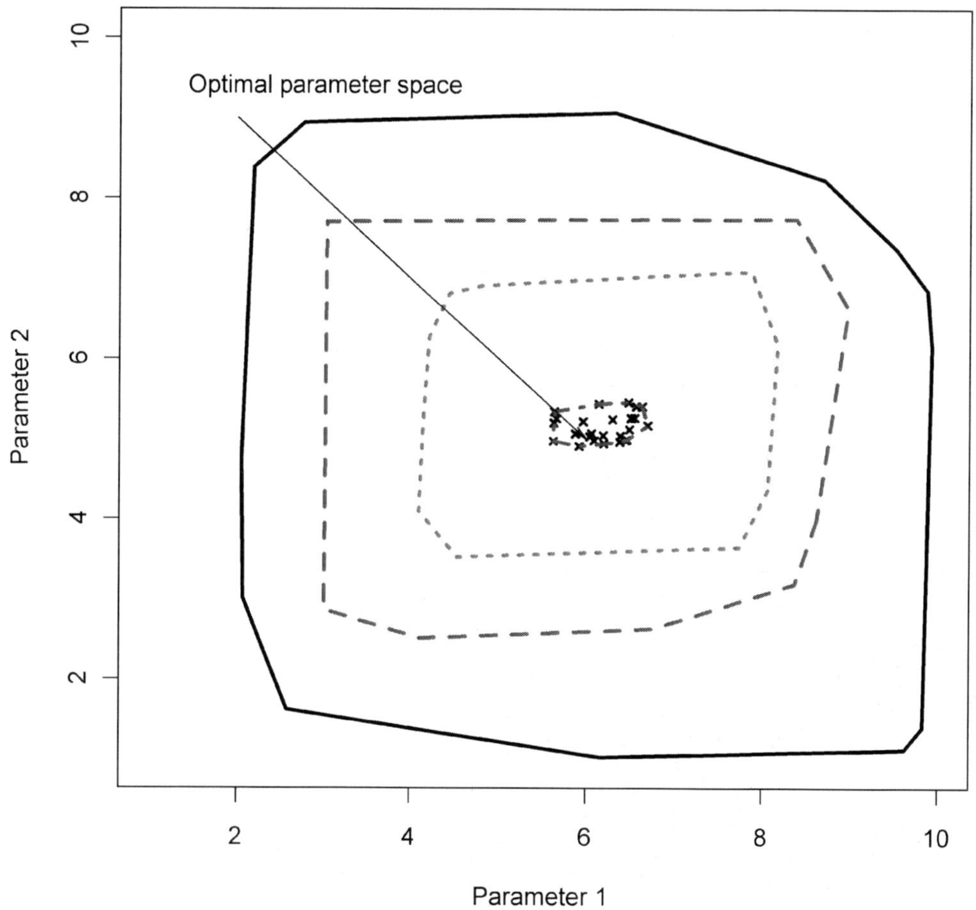

Figure 5. Systematic representation of ROPE algorithm.

A visual description of ROPE algorithm is given in Figure 5. From this figure we can see after several convex hull pealing we can reach an optimal parameter space. Bárdossy and Singh (2008) showed that optimal parameter space obtained by ROPE algorithm are good in transferability (transferring parameters for other time periods) and are less sensitive (small change in parameters set does not significantly change the performance of the model). Irrespective of other optimization algorithm, ROPE algorithm try to locate optimal parameter space, instead of optimal parameter sets. A single optimal parameter set is subset of optimal parameter space. Figure 6 shows an example of how ROPE algorithm goes toward optimal parameter sets. The X-axis of the figure is parameter set and Y-axis is objective function (NS). There can be several local minima and maxima. The basis aim of any global optmimization algorithm is to over come these local maimima or local minima and

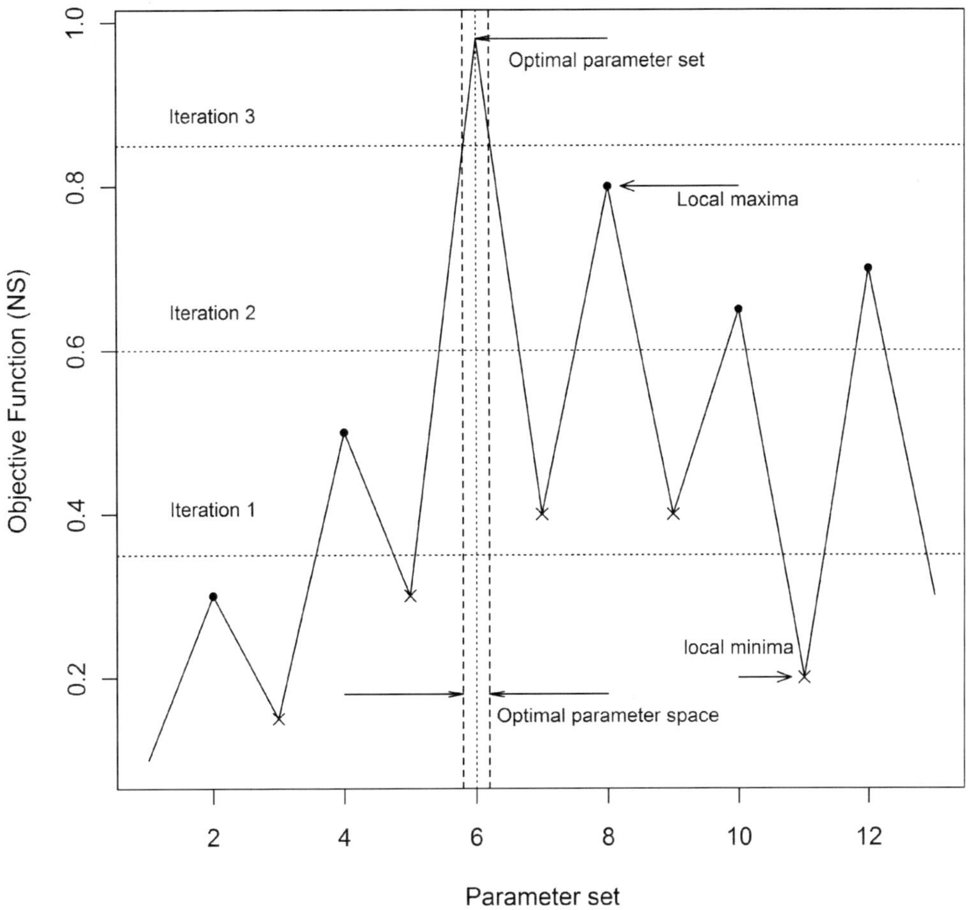

Figure 6. Objective function curve with respect to parameter sets-local minima and local maxima.

reach global optima. From this figure we can see after first iteration when we select the best parameter convex hull, we discards few local maxima and we move toward global maxima. Again in second iteration when we take convex hull of good parameters we discard few more local maxima and this process is repeated till we reach global optima.

2.3. Example of Calibration and Validation of a Hydrological Model

2.3.1. Study Area and Model

To explain the calibration and validation process in hydrological model, we applied the TopNet hydrological model in the Rangitata catchment located on South Island in New Zealand. The area of Rangitata catchment is 1468 km^2 and it receives an average of 2.32m of precipitation annually. The catchment elevation ranges from 0 to 2812m above mean sea level. The catchment has very high slope gradient. Meteorological data were obtained from Tait et al. (2006); Tait and Woods (2007) who interpolated data from over 500 climate stations in New Zealand across a regular 0.05ř latitude-longitude grid (approximately 5 km * 5 km). These data are provided at daily time steps, and are disaggregated to hourly data before use in the model.

TopNet is a semi-distributed physically based hydrological model which simulates catchment water balance and river flow. It was developed using the TOPMODEL concepts for parameterization of soil moisture deficit using a topographic index to model the dynamics of variable source areas contributing to saturation excess runoff (Bandaragoda et al., 2004). TopNet has two basic components (Clark et al., 2008)

- Simulating the water balance over a number of sub-catchments throughout a river basin
- Routing stream flow from each sub-catchments to the basin outlet

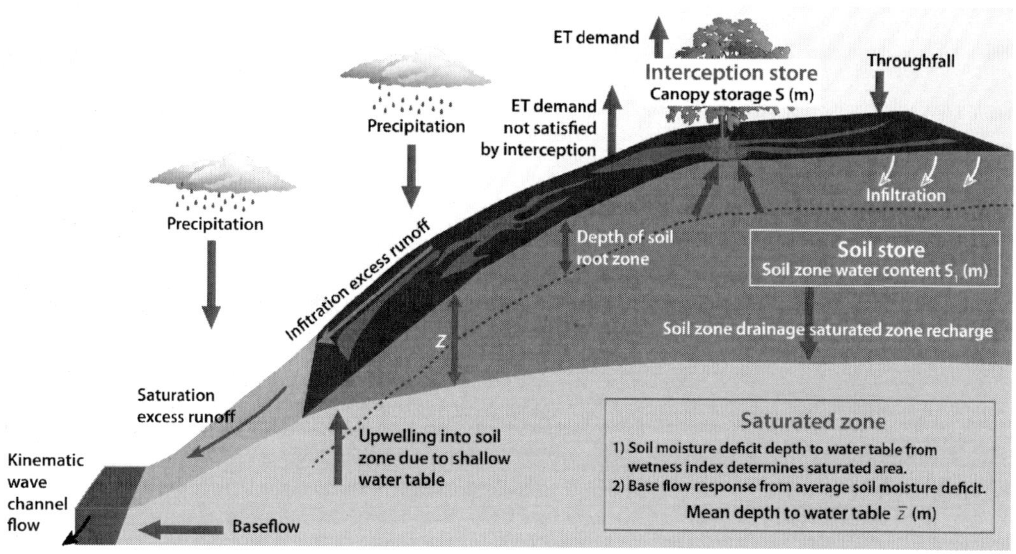

Figure 7. Systematic representation of Topnet Model (Bandaragoda et al., 2004).

TopNet can be used for a variety of applications including operational flood forecasting, water resources modelling, and climate and land-use change studies. ToNet model is widely used in real time flood forecasting in New Zealand. TopNet can provides a prediction of flow in each modelled reach within a catchment (Ibbitt et al., 2000; Bandaragoda et al., 2004; Clark et al., 2008). TopNet model needs time series of rainfall and temperature, plus other climate variables where available as a input. Topnet uses information on catchment topography, physical and hydrological properties, which is derived from a variety of sources, including a digital river network (River Environment Classification; Snelder and Biggs (2002), 30 m Digital Elevation Model, and land cover and soils databases (Land Cover Database; Land Resource Inventory; Newsome et al. (2000). Systematic representation of the model is given in Figure 7. TopNet has 31 parameters to define the hydrological

Table 2. TopNet model parameters of the which need calibration and possible range for the parameter multiplier

Parameter	Description	Initial Min	Range Max
topmodf	TOPMODEL f parameter	0.001	1
hydcon0	saturated hydraulic conductivity	0.01	9999
swater1	drainable water	0.05	20
swater2	plant-available water	0.05	20
dthetat	soil water content	0.1	10
overvel	overland flow velocity	0.1	10
gucatch	gauge under-catch for snowfall	0.5	1.5
th_{accm}	threshold for snow accumulation	272.16	275.16
th_{melt}	threshold for snow melt (K)	272.16	275.16
snowddf	mean degree-day factor for snow melt ($mmK^{-1}day^{-1}$)	0.1	7.5
minddfd	minimum degree-day-factor day (julian day: 1 to 366)	1	366
maxddfd	maximum degree-day-factor day (julian day: 1 to 366)	1	366
snowamp	seasonal amplitude of degree-day factor for snow melt ($mmK^{-1}day^{-1}$)	0	7.5
cv_{snow}	coefficient of variation in sub-grid SWE	0.5	1.5
r_{man_n}	Manning's n	0.1	10

processes of a catchment. Where possible, parameter values are determined from physical catchment properties; however 12 parameters typically require calibration. During calibration, TopNet model uses a spatially constant multiplier for each parameter, to adjust the parameters while retaining the relative spatial pattern obtained from the soil and vegetation data Bandaragoda et al. (2004). This procedure is necessary in order to reduce the dimensionality of the calibration problem. The model parameters which requires calibration for TopNet model and their physical meaning are given in Table 2.

2.4. Calibration and Validation

Proper choice of calibration data may play a great role in parameter identification of hydrological model. It is very difficult to say which data and what length of data is sufficient for the proper choice of calibration data which will give proper identification of model parameters. The data that contains lots of hydrological variability may be the best choice for calibration data because it may contain most of information that is required for parameter identification (Gupta and Sorooshian, 1985; Singh and Bárdossy, 2012; Singh et al., 2013). General practice is to use long series of data to calibrate the model and test the model on different set of data. In this chapter hourly data from year 2004 to 2006 is used for calibration and from year 2007 to 2009 for validation. It is very important to know in which parameter space we are searching the optimal parameters. Hence, first, a few model runs were taken to have some idea about the range of the parameters. A large number of uniformly distributed parameters sets were generated within the identified range and the model runs were taken with each set of parameters. The Nash-Sutcliffe (NS) coefficient (Nash and Sutcliffe, 1970) was used as the objective function for the model evaluation. Based on NS coefficient, the best 10 percent parameter sets were chosen and again the maximum and minimum values of the parameters were determined. After getting a proper range of parameters, the model was calibrated using the data for the period 2004-2007 using the ROPE algorithm. We obtained NS of 0.66.

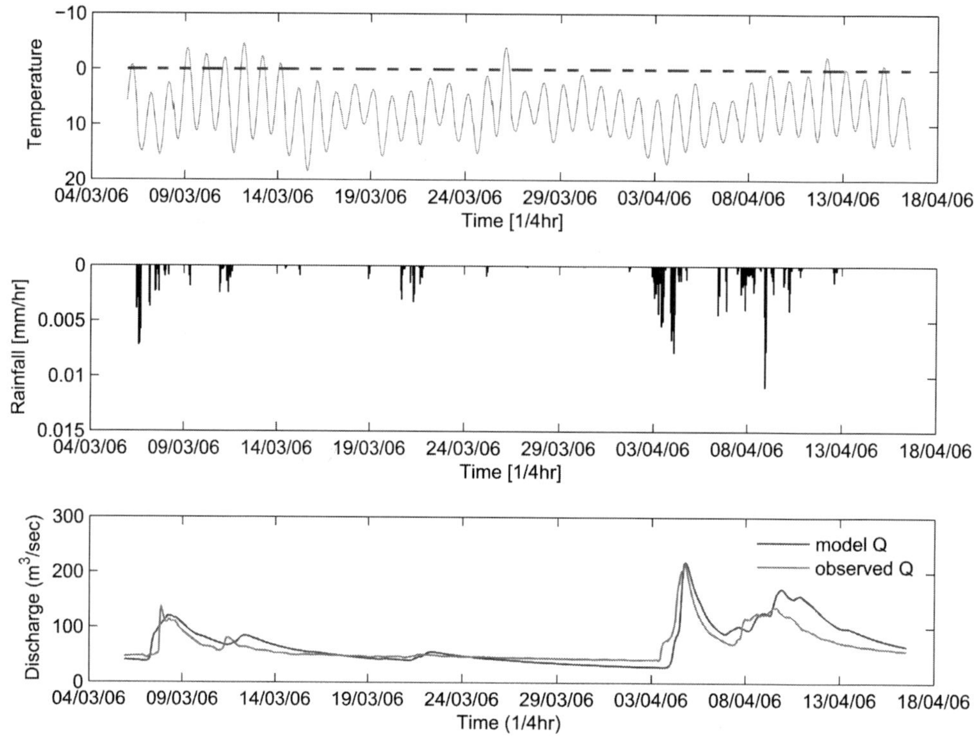

Figure 8. Hydrograph for calibration time period.

Figure 8 shows the model and the observed hydrograph for the calibration period. It can be seen that during the calibration runs, the catchment dynamics has been captured well at low peaks but the model has not given good results for very high peaks. This could be due to data limitations or model structure. During calibration, a very wide range of model parameters has been considered to avoid local minima. The ROPE algorithm has a feature that it does not give a single value of parameters after calibration; instead, it gives a vector of parameter set. After the calibration, the range of parameters has been considerably narrowed down. Figure 9 show the change is volume of parameters space with respect to the number of iteration in ROPE algorithm. It is clear from the figure that initial gradient of change in volume curve is very steep and then it have mild slope. This is simply because, once we are in optimal convex hull of the parameters, all the model parameter perform equally well. The model was validated using the data for the period 2007-2009. During this time period we obtained NS of 0.62. The value of NS coefficient is not very good but acceptable as compared to calibration NS. Figure 10 shows the observed and simulated hydrographs for the validation period. Further, instead of single value of each parameter, the ROPE calibration gives a space of good parameter sets. Hence, if desired ensemble of output can be generated. It has been found that the algorithm is effective in estimating parameters of a hydrological model.

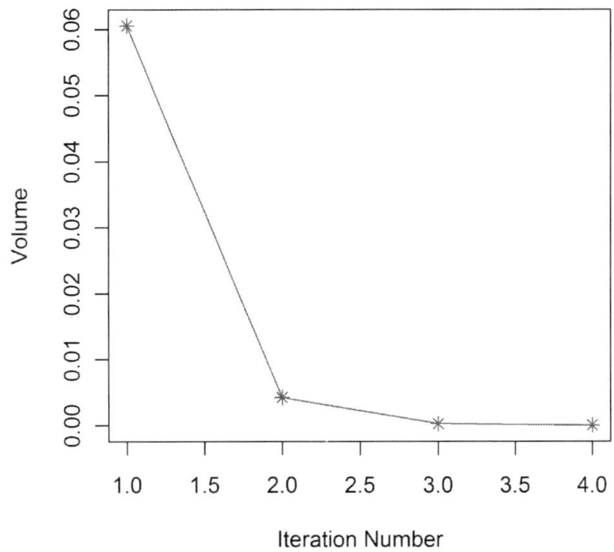

Figure 9. Change in volume of parameter space during calibration.

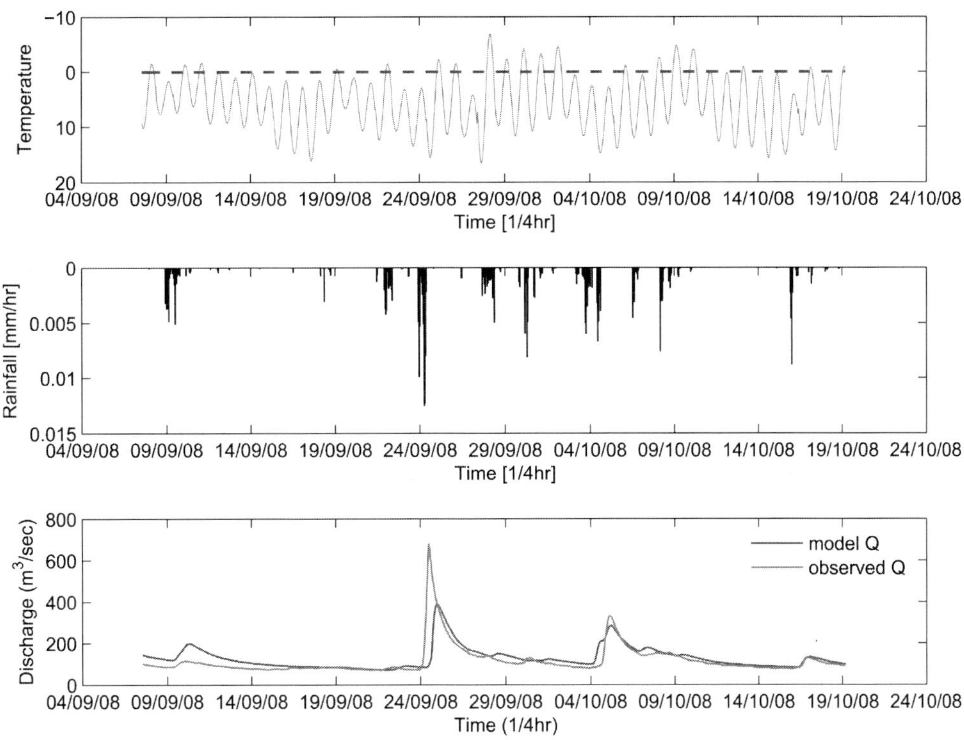

Figure 10. Hydrograph for validation time period.

3. Conclusion

The parameter estimation of any model is a difficult task. Reasons for this are the limitation of representation of processes in mathematical expression and then solving the mathematical expression. Also, the fact that different parameter vectors which describe the processes in the models might have the same effect on the output. This means that changes of some parameters might be compensated by others. Unfortunately, traditional manual calibration of models with reasonable parameter values often leads to weak results. Hence, nowadays, automatic procedures based on numerical methods are used. Many different optimization routines have been developed to find optimum parameter vectors. A variety of objective functions that measure model performance, including multi-objective approaches, have been tried to define optimality in this context. Nonlinearity of the hydrological models and of the objective functions lead to very complex optimization problems. Beven and Freer (2001) argue that there are no optimum parameters; in fact there is a large set of parameter vectors which all perform reasonably and one cannot easily distinguish between them. They call this an *equifinality* problem which leads to high uncertainties in the model predictions. Bárdossy (2007) investigated the geometrical properties of a parameter vector with good performance for a two-dimensional case and found that the set of good parameters was well structured. Unfortunately in higher dimensional spaces one can not see these sets, thus it is not clear whether they are scattered or have some clear structure. In this chapter the process of calibration and validation using hydrological model has been demonstrated. ROPE algorithm can not only be used for hydrological model but can be used for any kind of model. Concept presented in this chapter can be applied for any model.

References

Andréassian, V., Perrin, C., Michel, C., Usart-Sanchez, I., Lavabre, J., 2001. Impact of imperfect rainfall knowledge on the efficiency and the parameters of watershed models. *Journal of Hydrology* 250, 206–223.

Bandaragoda, C., Tarboton, D., Woods, R., 2004. Application of topnet in the distributed model intercomparison project. *Journal of Hydrology* 298 (1-4), 178–201.

Bárdossy, A., 2007. Calibration of hydrological model parameters for ungauged catchments. *Hydrol. Earth Syst. Sci.* 11, 703–710.

Bárdossy, A., Singh, S. K., 2008. Robust estimation of hydrological model parameters. *Hydrology and Earth System Science* 12, 1273–1283.

Bárdossy, A., Singh, S. K., 2008. Robust estimation of hydrological model parameters. Hydrology and Earth System Sciences 12 (6), 1273–1283.
URL `http://www.hydrol-earth-syst-sci.net/12/1273/2008/`

Bergström, S., Graham, L., 1998. On the scale problem in hydrological modelling. *Journal of Hydrology* 211 (1-4), 253–265.

Beven, K., 2000. Rainfall-Runoff Modelling: The Primer. Wiley.

Beven., K., 2006. A manifesto for the equifinality thesis. *Journal of Hydrology* 320, 18–36.

Beven, K., Freer, J., 2001. Equifinality, data assimilation, and data uncertainty estimation in mechanistic modelling of complex environmental systems using the glue methodology. *Journal of Hydrology* 249, 11–29.

Blasone, R. S., 2007. Parameter estimation and uncertainty assessment in hydrological modellingPh.D. dissertation, Technical University of Denmark.

Blöschl, G., Sivapalan, M., 1995. Scale issues in hydrological modelling: a review. *Hydrological processes* 9 (3-4), 251–290.

Clark, M. P., Rupp, D. E., Woods, R. A., Zheng, X., Ibbitt, R. P., Slater, A. G., Schmidt, J., Uddstrom, M. J., 2008. Hydrological data assimilation with the ensemble kalman filter: Use of streamflow observations to update states in a distributed hydrological model. Advances in Water Resources 31 (10), 1309–1324.

Donoho, D. L., Gasko, M., 1992. Breakdown properties of location estimates based on halfspace depth and projected outlyingness. *The Annals of Statistics 20(4), 1808–1827.*

Gupta, H., Sorooshian, S., Hogue, T. S., Boyle, D. P., 2003. Advances in calibration of watershed models. Calibration of Watershed Models. AGU Monograph Series. *Water Science and Application* 6, 9–28.

Gupta, H., Sorooshian, S., Yapo, P., 1998. Toward improved calibration of hydrologic models: Multiple and noncommensurable measures of information. *Water Resources Research* 34 (4), 751–763.

Gupta, V. K., Sorooshian, S., 1985. The relationship between data and the precision of parameter estimates of hydrologic models. Journal of Hydrology 81, 57–77.

Hendrickson, J. D., Sorooshian, S., Brazil, L. E., 1988. Comparison of newton-type and direct search algorithms for calibration of conceptual rainfall-runoff models. *Water Resources Research* 24 (5), 691–700.

Ibbitt, R. P., Henderson, R. D., Copeland, J., Wratt, D. S., 2000. Simulating mountain runoff with meso-scale weather model rainfall estimates: a new zealand experience. *Journal of Hydrology* 239 (1âĂŞ4), 19–32.

Jackson, D. R., Aron, G., 1971. Parameter estimation in hydrology: the state of the art. *Water Resources Bulletin* 7 (3), 457–472.

Jakeman, A. J., Hornberger, G. M., 1993. How much complexity is warranted in a rainfall-runoff model? *Water Resources Research* 29 (8), 2637–2650.

Kavetski, D., Kuczera, G., Franks, S., 2006a. Calibration of conceptual hydrological models revisited: 2. improving optimisation and analysis. *Journal of Hydrology* 320, 187–201.

Kavetski, D., Kuczera, G., Franks, S. W., 2006b. Calibration of conceptual hydrological models revisited: 1. overcoming numerical artefacts. *Journal of Hydrology* 320, 173–186.

Kuczera, G., 1997. Efficient subspace probabilistic parameter optimization for catchment model. *Water Resources Research* 33 (1), 177–185.

Liu, R. Y., 1990. On a notion of data depth based on random simplices. *The Annals of Statistics* 18, 405–414.

Liu, R. Y., Parelius, J. M., Singh, K., 1999. Multivariate analysis by data depth: descriptive statistics, graphics and inference. *The Annals of Statistics* 27(3), 783–858.

Michaud, J., Sorooshian, S., 1994. Comparison of simple versus complex distributed runoff models on a midsized semiarid watershed. *Water Resources Research* 30 (3), 593–605.

Moradkhani, H., Sorooshian, S., Gupta, H. V., Houser, P. R., 2005. Dual state-parameter estimation of hydrological models using ensemble kalman filter. *Advances in Water Resources* 28 (2), 135 – 147.

Nash, J., Sutcliffe, J., 1970. River flow forecasting through conceptual models. 1. a discussion of principles. *Journal of Hydrology* 10, 282–290.

Newsome, P., Wilde, R., Willoughby, E., 2000. Land resource information system spatial data layers. *Landcare Research Technical Report* No.84p.

Patil, S. R., 2008. Regionalization of an event based nash cascade model for flood prediction in ungauged basinsPh.D. dissertation No. 175, University of Stuttgart.

Perrin, C., Michel, C., Andréassian, V., 2001. Does a large number of parameters enhance model performance? comparative assessment of common catchment model structures on 429 catchments. *Journal of Hydrology* 242 (3-4), 275–301.

Perrin, C., Michel, C., Andréassian, V., 2003. Improvement of a parsimonious model for streamflow simulation. *Journal of Hydrology* 279 (1-4), 275–289.

Refsgaard, J., Knudsen, J., 1996. Operational validation and intercomparison of different types of hydrological models. *Water Resources Research* 32 (7), 2189–2202.

Rousseeuw, P. J., Struyf, A., 1998. Computing location depth and regression depth in higher dimensions. *Statistics and Computing* 8, 193–203.

Samaniego, L., Bárdossy, A., 2005. Robust parametric models of runoff characteristics at the mesoscale. *Journal of Hydrology* 303, 136–151.

Singh, S., Bárdossy, A., Götzinger, J., Sudheer, K., 2012a. Effect of spatial resolution on regionalization of hydrological model parameters. *Hydrological Processes* 26, 3499–3509.

Singh, S. K., 2010. Robust parameter estimation in gauged and ungauged basins.Ph.D. dissertation No. 198, University of Stuttgart.

Singh, S. K., Bárdossy, A., 2012. Calibration of hydrological models on hydrologically unusual events. *Advances in Water Resources* 38, 81–91.

Singh, S. K., Jiaying, L., Bárdossy, A., 2012b. Improving calibration strategy of physically based model wasim-eth using critical events. *Hydrological Sciences Journal* 57 (8), 1487Ű1505.

Singh, S. K., McMillan, H., Bárdossy, A., 2013. Use of the data depth function to differentiate between case of interpolation and extrapolation in hydrological model prediction. *Journal of Hydrology* 477, 213Ű228.

Singh, V. P., Woolhiser, D. A., 2002. Mathematical modeling of watershed hydrology. *Journal of Hydrologic Engineering* 7 (4), 270–292.

Sivapalan, M., Takeuchi, K., Franks, S. W., Gupta, V. K., Karambiri, H., Lakshim, V., Liang, X., McDonnell, J. J., Mendiondo, E. M., Connell, O., Oki, T., Pomeroy, J., Schertzer, D., Uhlenbrook, S., Zehe, E., 2003. Iahs decade on predictions in ungauged basins (pub), 2003-2012: Shaping an exciting future for the hydrological sciences. *Hydrological Sciences Journal* 48 (6), 857–880.

Snelder, T., Biggs, B., 2002. Multiscale river environment classification for water resources managements1. *JAWRA Journal of the American Water Resources Association* 38 (5), 1225–1239.

Sorooshian, S., Duan, Q., Gupta, V. K., 1993. Calibration of rainfall-runoff models: application of global optimization to the sacramento soil moisture accounting model. *Water Resources Research* 29 (4), 1185–1194.

Tait, A., Henderson, R., Turner, R., Zheng, X., 2006. Thin plate smoothing spline interpolation of daily rainfall for new zealand using a climatological rainfall surface. *International Journal of Climatology* 26 (14), 2097–2115.

Tait, A., Woods, R., 2007. Spatial interpolation of daily potential evapotranspiration for new zealand using a spline model. *Journal of Hydrometeorology 8 (3), 430–438.*

Tukey, J., 1975. Mathematics and picturing data. In: In *Proceedings of the 1975 International 17 Congress of Mathematics*. Vol. 2. pp. 523–531.

van der Linden, S., Woo, M., 2003. Application of hydrological models with increasing complexity to subarctic catchments. *Journal of Hydrology* 270 (1-2), 145–157.

Vrugt, J. A., Bouten, W., Gupta, H. V., Sorooshian, S., 2002. Toward improved identifiability of hydrologic model parameters: The information content of experimental data. *Water Resources Research* 38(12), 1312.

Wagener, T., McIntyre, N., Lees, M. J., Wheater, H. S., Gupta, H. V., 2003. Towards reduced uncertainty in conceptual rainfall-runoff modelling: Dynamic identifiability analysis. *Hydrological Processes* 17(2), 455Ű476.

Wagener, T., Wheater, H. S., Gupta, H. V., 2004. Rainfall-Runoff Modelling in Gauged and Ungauged Catchments. Imperial College Press, London.

Wikipedia, 2012. Convex hull. http://en.wikipedia.org/wiki/Convexhull.

WolframMathWorld, 2012. Convex hull. http://mathworld.wolfram.com/ ConvexHull.html.

Zuo, Y., Serfling, R., 2000. General notions of statistical depth function. *The Annals of Statistics* 28(2), 461–482.

In: Calibration Technology, Theories and Applications ISBN: 978-1-62618-808-2
Editors: Ikumatsu Fujimoto and Kunitoshi Nishimura

Chapter 7

CALIBRATION TRANSFER OF PLS MODELS USING VIRTUAL STANDARDS

Mohamed F. Abdelkader[1], John B. Cooper[1] and Christopher M. Larkin[2]

[1]Department of Chemistry, Old Dominion University, Norfolk, VA, US
[2]Bruker Optics, The Woodlands, TX, US

ABSTRACT

For chemical samples which involve highly complex mixtures, it is often difficult, if not impossible to maintain and transport physical standards for calibration transfer of spectral multivariate calibration models. This situation often arises due to sample stability, volatility, or complexity in preparation. This is particularly true in the fuel industry where thousands of unique chemicals often comprise a sample. In this chapter we provide detailed applications of the use of virtual standards in transferring partial least squares (PLS) calibration models which relate fuel properties to NIR spectra. The virtual standards are digitally constructed using only a few neat chemicals acquired on the master and slave instruments. The ability to acquire the neat chemicals at future times and geographic locations as well as the digital "mixing" procedure eliminates the problems described above. Calibration transfer using segmented virtual standards slope and bias (SVSSB) is demonstrated between different dispersive NIR instruments as well as between FT NIR and dispersive NIR instruments.

Keywords: Virtual Standards, PLS, SVSSB, Calibration Transfer

ABBREVIATIONS

SVSSB	Segmented Virtual Standards Slope and Bias
SVD	Singular Value Decomposition
SB	Slope and Bias Using Real Fuels
SNV	Standard Normal Variate

PLS	Partial Least Square
SEP	Standard error of prediction
SECV	standard error of cross validation
RMSEP	root mean square error of prediction
NIR	near infra-red

7.1. INTRODUCTION

Multivariate models have gained widespread acceptance as a secondary analytical method in fuel analysis, pharmaceutical industry, and food industry. These methods are typically based on a partial least squares (PLS) regression correlating a property of interest with near infrared (NIR) or Raman spectra. Although PLS models can be built in a laboratory environment using a spectrometer and a set of calibration samples, development and validation of a robust multivariate model often involves hundreds of samples and the use of elaborate instrumentation to determine reference values. This leads to a large investment in time and money. Therefore, it is beneficial to be able to use a validated PLS model built on a primary instrument on additional instruments or on the same instrument as changes occur to the instrument over time. However, when measuring a sample on another instrument or the same instrument after some time, changes in the instrumental response lead to erroneous predictions by PLS models. One way to solve this problem is to re-measure every sample and reconstruct the PLS model. Yet, this is an impractical solution since it is time consuming, expensive, and requires a significant effort. For this reason, there has been extensive work to develop multivariate models calibration transfer methods between instruments. Where the term “Calibration” here means making an old instrument or a newly-built instrument give the same reading as the instrument in which the PLS model was built. These calibration transfer methods can be generally characterized as either spectral transfer or model transfer methods.

In the case of spectral transfer, the response of a sample measured on the secondary instrument is corrected to the response obtained on the primary instrument so that the spectra of standards collected on the master and secondary instruments match each other. Spectral transfer methods include, but are not limited to, direct standardization, piecewise direct standardization, and orthogonal signal correction. On the other hand, in the case of model transfer, the PLS model built and validated on the primary instrument is adjusted in a way that makes the predicted values for the secondary instrument match the predicted values of the primary instrument. Examples of model transfer methods are simple slope and bias correction, model updating, and regression equation transfer. However, for both spectral and model transfer, there is a requisite to measure a number of samples (the transfer set) on both the primary and a secondary instrument. The transfer set spectra should closely match the spectra of the samples used to build the PLS model and must exhibit sufficient variability to account for the difference between the two instruments. Maintaining transfer set integrity is a challenge since contamination and compositional change can happen during transportation or storage over time due to component volatility and reactivity. This is particularly true with fuels which are composed of thousands of chemicals. If a viable transfer set cannot be maintained over time, then the master instrument must be maintained in pristine condition: an equally challenging task. In order to overcome these limitations, a new model transfer method is described herein, involving the use of spectra of pure molecular chemicals to digitally generate segmented spectra of virtual standards that mimic the training set spectra used to build the PLS models. Virtual standard spectra from both the primary and a secondary instrument are then predicted on the PLS models and a simple slope and bias correction can be applied to the secondary instrument.

7.2. Instrumentation

Two FUELex NIR spectrometers (BrukerOptics, The Woodlands, TX, www. brokeroptics. com\fuelex) were used to collect the NIR spectra of fuels and chemicals. The FUELex consists of an f#/2 spectrograph with a 512 element InGaAs array detector and 3 nm resolution. The spectral wavelength range is 880 - 1570 nm. Each spectral acquisition consisted of 30 averaged spectra each with an integration time between 8 to 16 ms. the integration time for each instrument was automatically selected so that the signal intensity of an air reference yielded between 32,000 and 42,000 counts on the 16 bit A/D converter. This ensured that sufficient signal was present while preventing the A/D from saturating when the probe was inserted into the fuel. Total acquisition time for each spectrum was less than one second. All fuel spectra exhibited absorbance ranges within 0-0.8 absorbance units.

A MPA NIR FT instrument (BrukerOptics, The Woodlands, TX, www.brukeroptics. com\MPA) was used to collect the NIR spectra of fuels and chemicals. The spectral wavenumber range is 4000 - 12000 cm^{-1}.

Each instrument was equipped with a transflectance dip probe with a fixed total pathlength of 1 cm. All spectra were collected at room temperature (22-25 C^{o}) by immersing a clean probe several inches into the sample and then swirling the probe to ensure no air bubbles were trapped. No effort was made to maintain a set fiber-optic geometry. All of the fuel and chemical spectra were preprocessed by Savitsky-Golay 7-point first derivative followed by a standard normal variate (SNV) transform to remove spectral off-set and baseline changes due to the variability of fiber-optic geometry in both Fuelex and MPA instruments. All chemicals were analytical grade (Aldrich).

7.3. Building PLS Models

NIR spectra of 420 JP-8 jet fuels were acquired using the primary instrument. The fuel samples were obtained from the Naval Research Laboratory in Washington D.C and from the Army TARDEC facility in Warren Michigan. The initial origin of the fuels ranged geographically on a sample by sample basis throughout the North American, European, African, Australian and Asian continents. World-wide representation of fuel samples ensured variance in the training set used for PLS model building. ASTM reference values for each fuel set were provided; yet, most fuels lacked reference values for all the three properties listed in Table 1. ChemWave software (Bruker Optics, Woodlands, TX, USA) was used to construct and validate all PLS models of jet fuel properties from the spectra acquired on the primary instrument. Constructing a PLS model involves the following steps;

1. Selecting fuel spectra whose ASTM number for the property of interest are known, and combining these spectra with their corresponding property values into a training set,
2. Constructing a PLS model which correlates spectral intensity of the property of interest,
3. Selecting the number of principal components (latent variable) to use in the model,
4. Identifying and removing any outlier values,
5. Validating PLS model using leave-one-out cross validation,
6. Validating PLS model with an independent test set.

Table 1. Summary of the Jet Fuel PLS Models and Properties

Jet fuel property	ASTM method*	Units	Number of samples	ASTM Reproducibility	$SECV_M$	Mean	Std. Dev.	r^2
API Gravity	D1298	°API	291	0.3	0.265	43.94	3.38	0.98
Cetane Index	D976	Cetane #	138	2	0.309	45.83	1.72	0.97
Hydrogen Content	D3343	% (mass)	114	0.04	0.019	13.82	0.13	0.97

* http://www.astm.org/

$SECV_M$: Standard error of validation of primary instrument property model using leave-one-out validation; Mean: mean of the training set values; Std. Dev.: standard deviation of the training set values; r^2: correlation coefficient for predicted vs. actual values in leave-one-out validation plots. See Table 1 for units of measurement.

The number of latent variables was determined using the IND function and an F-test with (n-k) degree of freedom and a probability level of 95%. For all the three property models shown in this chapter, five latent variables were used. Five latent variables were appropriate and showed the lowest SECV (standard error of cross-validation) with the minimum over-fitting of noise. Samples were considered outliers if both studentized residuals and leverage exceeded 98% probability of training set inclusion. Sample leverage is calculated using eqn 1

$$\text{sample leverage} = [\textit{scores}\ (\text{SCORES X SCORES}^{T})\ \textit{scores}^{T}]/n \qquad (1)$$

where *SCORES* is a matrix of the entire training set where each row is a sample spectrum and numbers of columns give the latent variables used and $SCORES^T$ is the transposed matrix, and where *scores* is a row vector containing the amount of each latent variable, and $scores^T$ is the transposed vector, and n is the number of samples in the training set.

In step 5, a particular model was validated by leave-one-out cross validation where one sample is removed from the training set, and then a PLS model was built using the remaining samples and finally this model is used to predict the property value of the excluded sample. This process is repeated until all the samples in the training set are excluded, each one once. The standard error of cross validation (SECV) is then calculated as in eqn 2

$$\text{SECV} = [(\Sigma(\text{error})^2)/(n\text{-}k)]^{1/2} \qquad (2)$$

where error is the difference between the true value and the predicted value for all leave-one-out iterations, n is the total number of samples used to build the model, and k is the number of latent variable used to generate a model.

Table 2. Standard Error of Prediction of the properties in the test set samples relative to the master instrument prediction for uncorrected and SVSSB corrected

Jet fuel property	Samples Number	Primary Instrument	Secondary Instrument			
		SEP	RMSEPu	RMSEPc	SB	r2
API Gravity	180	0.32	1.46	0.18	0.18	0.93
Cetane Index	180	0.67	1.09	0.23	0.27	0.98
Hydr. Cont.	180	0.05	0.057	0.015	0.014	0.98

SEP: Standard Error of Prediction of primary instrument relative to the actual property values; RMSEPu: Root Mean Square Error of Prediction of uncorrected secondary instrument relative to primary instrument prediction; RMSEPc: Root Mean Square Error of Prediction of SVSSB corrected secondary instrument relative to primary instrument prediction; r2: correlation coefficient of primary prediction relative to secondary prediction.

The ASTM reproducibility, the standard error of cross-validation, the mean and the standard deviation of actual values, and the correlation coefficient for the predicted vs. actual results are given in Table 1. ASTM reproducibility values reflect the maximum expected inter-lab measurements error for the primary ASTM test method at a 95% confidence level (2.6 times the standard deviation of the ASTM measured population). The standard error of cross validation (SECVM) for the 3 PLS models of jet fuels properties are within the ASTM primary methods. The standard deviations of the 3 jet fuel properties reflect the global nature of the jet fuel samples used in building these PLS models and the correlation coefficients values indicate that all the samples used are linearly correlated.

NIR spectra of an additional 180 JP-8 jet fuels (TARDEC, Warren, MI, USA) were acquired on both the primary and secondary instruments to form an independent test set (step 6). The independent test acquired on the primary instrument was used to validate the PLS models. The standard errors of prediction (SEP) are given in Table 2.

7.4. Segment Virtual Standard Calibration Transfer

7.4.1. Theory

A virtual standard that mimics the actual training set acquired on a master instrument, and that is used to build PLS models, can be digitally generated by using a few spectra of neat molecular chemicals acquired on the master instrument, the spectra of the training set, and a multiple linear regression algorithm. By acquiring the spectra of these neat molecular chemicals on a secondary instrument, virtual standards that represent the change on the secondary instrument can be digitally generated. Finally, a simple but effective slope and bias correction can be applied to the secondary instrument to correct PLS models predictions. A schematic summary of the segmented virtual standards slope-bias (SVSSB) method is given in Figure 1. To generate segmented virtual standards we used 15 neat molecular chemicals. The first derivative SNV spectra of the fifteen pure chemicals (butylcyclohexane, cis-decahydronaphtalene, dodecane, ethylebenzene, hexadecane, iso-octane, meta-xylene, nonane, ortho-xylene, pristane, squalene, tetradecane, toluene, tridecane, and unidecane) are shown in Figure 2. As shown in Figure 1, the target spectra (the original training set fuel spectra acquired on the primary instrument) are divided into three segments. Segment one is from 880 to 1304 nm, segment two is from 1304 to 1432 nm and segment three is from 1432 to 1570 nm. This process is repeated for the 15 pure chemicals acquired on the primary instrument.

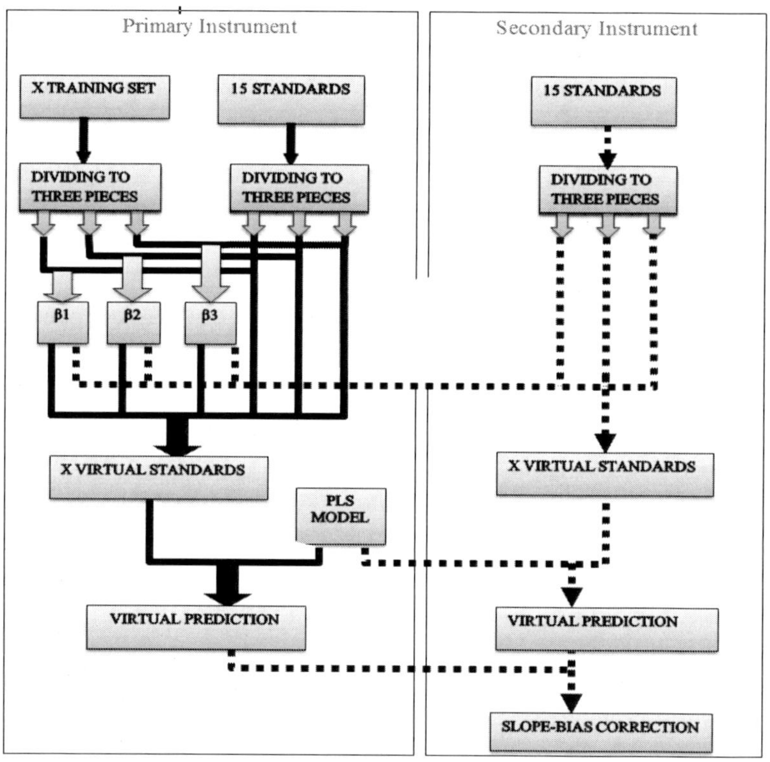

Figure 1. Calibration transfer using segmented virtual standards which are digitally synthesized using 15 pure chemicals.

$$f_{target} = [\mathbf{f}_1..\mathbf{f}_j] \tag{3}$$

$$x_{chemical} = [\mathbf{x}_1..\mathbf{x}_j] \tag{4}$$

where f_{target} is a fuel spectrum and $\mathbf{f}_j$ is a segment of this spectrum and $\mathbf{x}_{chemical}$ is a chemical spectrum (toluene, nonane,... etc) and $\mathbf{x}_j$ is a segment of this spectrum. The brackets in equations 1 and 2 represent the vector concatenation operator. For each target fuel, the three segments are then approximated (one at a time) using a linear combination of the corresponding segments of the 15 pure chemicals. This was accomplished by first setting up a multiple linear regression:

$$f_j = \beta_{1j} x_{1j} + \beta_{2j} x_{2j} + \ldots \beta_{ij} x_{ij} + \varepsilon \quad (5)$$

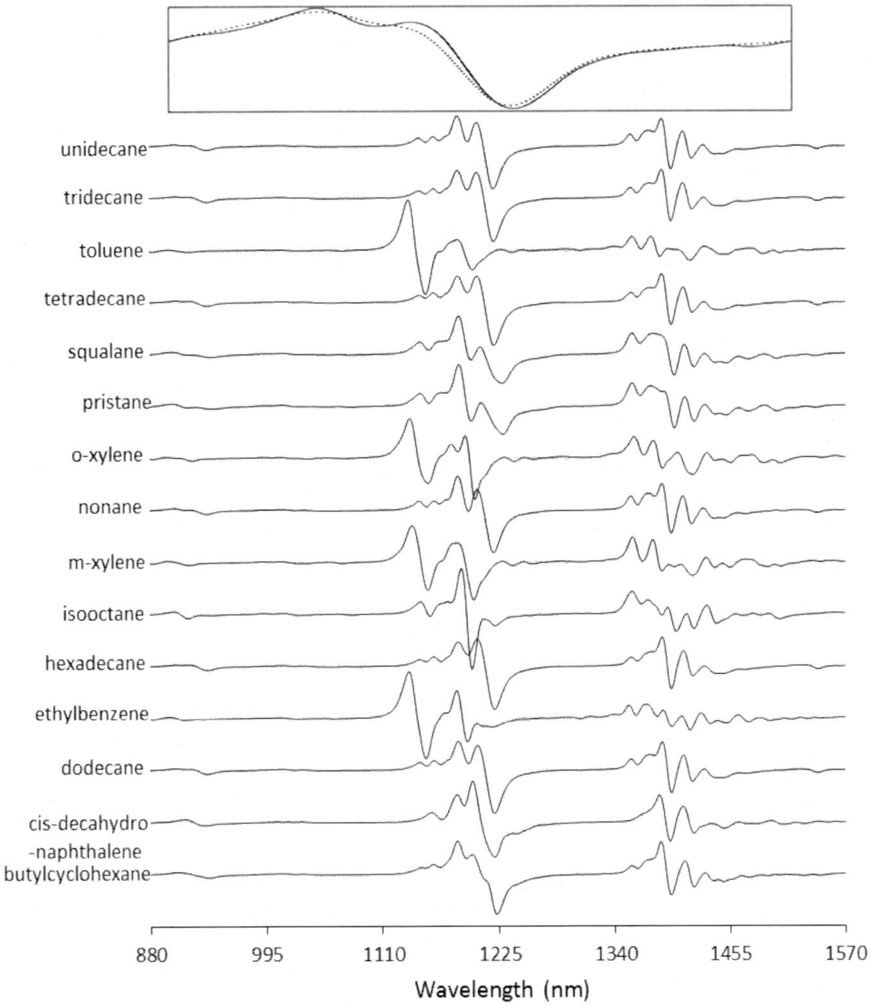

Figure 2. NIR absorbance of 1st derivative SNV spectra of the 15 chemicals acquired on the primary instrument (Solid line) overlaid with the secondary standards (broken line). Seven point Savitsky-Golay 1st derivative and SNV spectral transforms have been applied. Spectra offset by 6 a. u.

where f_j is a segment of a virtual fuel standard which approximates a particular target fuel; $\boldsymbol{x}_{ij}$ is a segment j of a chemical spectrum i; i is the chemical spectrum index (1-15); j is the spectrum segment index (1-3); and $\boldsymbol{\beta}_{ij}$ is the amount of each chemical ($\boldsymbol{x}_{ij}$) required to make segment j of the virtual standard. The remaining term $\boldsymbol{\varepsilon}$ is the spectral difference between the virtual standard segment and the corresponding segment of the target fuel being approximated. By applying eqn. 5 to all three segments sequentially, and then concatenating the results, a virtual standard is formed which approximates one of the target fuels. Using matrix notation, eqn. 3 can be re-written to solve this relationship for every target fuel spectrum to be digitally synthesized:

$$F_j = X_j\beta_j + \varepsilon \tag{6}$$

where, $\mathbf{F_j}$ is a n x m matrix where the number of rows (n) corresponds to the number of wavelengths in the spectral segment *j*, and the number of columns (m) corresponds to the number of target fuels being approximated. Thus each column of $\mathbf{F}_j$ is segment *j* of a virtual standard. The matrix $\mathbf{X_j}$ is an n x i matrix where the number of rows (n) also corresponds to the number of wavelengths in the spectral segment *j*, and the number of columns (i) corresponds to the number of pure chemicals (15). Thus each column is segment *j* of chemical spectrum *i*. $\boldsymbol{\beta_j}$ is a matrix relating how much of each chemical is required to generate the targets for segment *j*. By setting the spectral difference matrix ($\boldsymbol{\varepsilon}$) to zero, the matrix $\boldsymbol{\beta_j}$ is solved as:

$$\beta_j = (X_j^T X_j)^{-1} X_j^T F_j \tag{7}$$

using singular value decomposition. In this chapter, the SVD was performed at full rank (15), thus conceptually the same as a multiple linear regression. The virtual standards for a particular segment *j* are then generated as:

$$S_j = X_j \, \beta_j \tag{8}$$

The complete set of virtual standards is then given by:

$$S = [S_1..S_j] \tag{9}$$

where the brackets indicate the matrix concatenation operator and each column of S is a virtual standard approximating a target fuel acquired on the primary instrument. For chemical spectra also acquired on the primary instrument, S becomes $S_{primary}$ (the virtual standard spectra for the primary instrument).

An example of a virtual standard generated using this method is given in Figure 3, and is overlaid with both the target fuel spectrum and a virtual standard created without using the segmented approach. Once $\boldsymbol{\beta_j}$ is determined, then chemical spectra acquired on the secondary instruments are used to generate the virtual standards for each secondary instrument:

$$S_{secondary} = [S_{1,secondary} \ldots S_{j,secondary}] = [X_{1,secondary}\beta_1 \ldots X_{j,secondary}\beta_j] \tag{10}$$

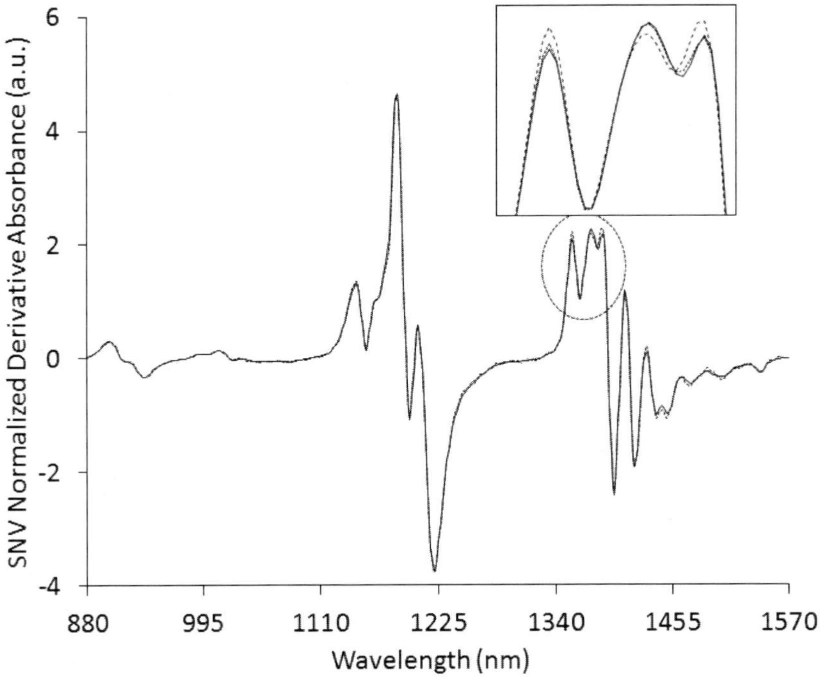

Figure 3. NIR absorbance spectrum of a digitally synthesized virtual standard using segment target (dotted line) overlaid with digitally synthesized virtual standard using full target (dashed line) and their target fuel spectrum (solid line). Seven point Savitsky-Golay 1st derivative and SNV spectral transforms have been applied.

where this equation is the same as equation 5, except the chemical spectra in the $\mathbf{X}_{j,secondary}$ matrix are acquired on the secondary instrument instead of the primary instrument.

The virtual standards of the primary instrument ($S_{primary}$) and the secondary instrument ($S_{secondary}$) are then predicted in a PLS model. Subsequently, a slope and bias correction for each secondary instrument PLS prediction is generated by regressing the predictions of the primary instrument virtual standards against those of the secondary instrument virtual standards.

As an example, when correcting the primary instrument generated PLS model for API gravity, the primary instrument virtual standards ($S_{primary}$) are predicted using the PLS model to yield $P_{primary}$, and the secondary instrument virtual standards ($S_{secondary}$) are predicted using the PLS model to yield $P_{secondary}$. The slope-bias correction for future API gravity predictions on the secondary instrument is then given by solving:

$$P_{primary} = a * P_{secondary} + b \tag{11}$$

for the slope (a) and the bias (b). Future predictions on the secondary instrument are simply corrected as:

$$P_{corrected} = a * P_{secondary} + b \tag{12}$$

The essence of the method is that once the $\boldsymbol{\beta}_j$ values have been determined for the primary instrument, transfers can be carried out to an unlimited number of secondary instruments with only 1 requirement: the acquisition of the 15 pure chemicals on the secondary instrument. Since each chemical has a discrete molecular formula, they can always be obtained in the future in pristine condition. There is also no longer a need to maintain the primary instrument once the 15 chemical spectra have been acquired on it.

7.4.2. SVSSB Calibration Transfer between Similar NIR-Instruments

NIR spectra of 180 JP-8 jet fuels (TARDEC, Warren, MI, USA) were acquired on both the primary and secondary instruments to form an independent test set. The standard errors of predictions on the primary and secondary instruments are given in table 2. As shown in table 2, the root mean square error of prediction ($RMSEP_u$) for the secondary instrument relative to the primary instrument is far from SEP of the primary instrument and fall outside ASTM reproducibility for all of the 3 jet fuel properties tested in this chapter; API, cetane Index, and % hydrogen content.

To correct the prediction of the secondary instrument, the 15 neat chemicals, listed above and shown in Figure 2, were acquired on both the primary instrument and secondary instruments. Even though, the two instruments were of the same type, the spectra of the 15 neat chemicals were slightly different as shown in the inset of Figure 2. The spectra of the 15 neat chemicals, acquired on both primary and secondary instruments, and the spectra of the training set used in building PLS models were used to generate virtual standards for both primary and secondary instruments as described in the theory section. The generated virtual standards of the primary and secondary instrument were regressed against each other and a slope and bias correction was applied to the secondary instrument. The root mean square error of prediction ($RMSEP_C$) for the secondary instrument after SVSSB correction is shown in Table 2. After applying SVSSB correction the $RMSEP_C$ for the secondary instrument relative to the primary instrument is within ASTM reproducibility.

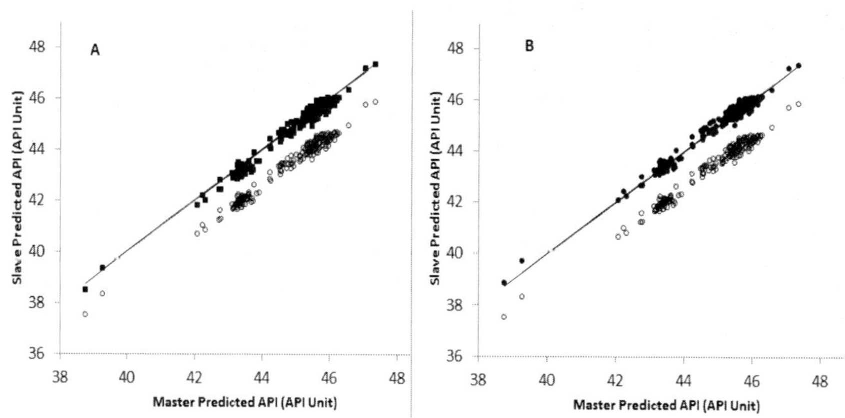

Figure 4. API predictions of test set fuels acquired on the secondary instrument for uncorrected (open circles), SVSSB corrected (solid squares) and BS corrected (solid circles) data. (A) Shows SVSSB correction and (B) shows SB correction using actual fuel standard. The solid line is an ideal line (slope=1 and intercept =0).

To test the reliability of the SVSSB method, the training set samples were acquired on the secondary instrument and a slope and bias correction was carried out using real fuel spectra of both training sets acquired on the primary and the secondary instruments. The root mean square error of prediction of the secondary instrument relative to primary instrument using real fuel correction (SB) is shown in table 2 and they are similar or close to the prediction in the case of SVSSB correction.

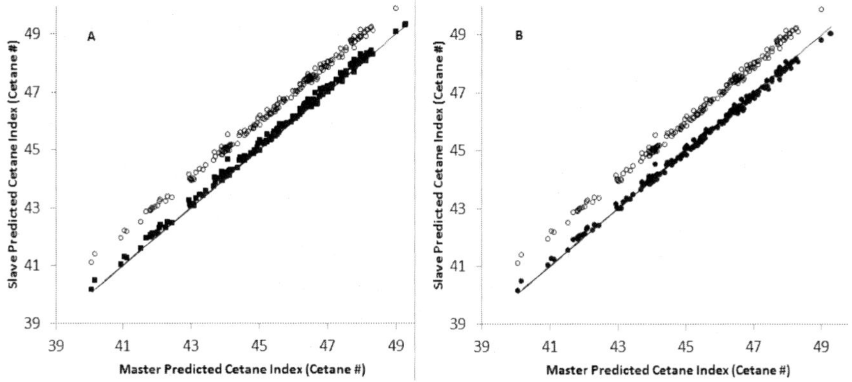

Figure 5. Cetane Index predictions of test set fuels acquired on the secondary instrument for incorrected (open circles), SVSSB corrected (solid squares) and BS corrected (solid circles) data. (A) Shows SVSSB correction and (B) shows SB correction using actual fuel standards. The solid line is an ideal line (slope=1 and intercept =0).

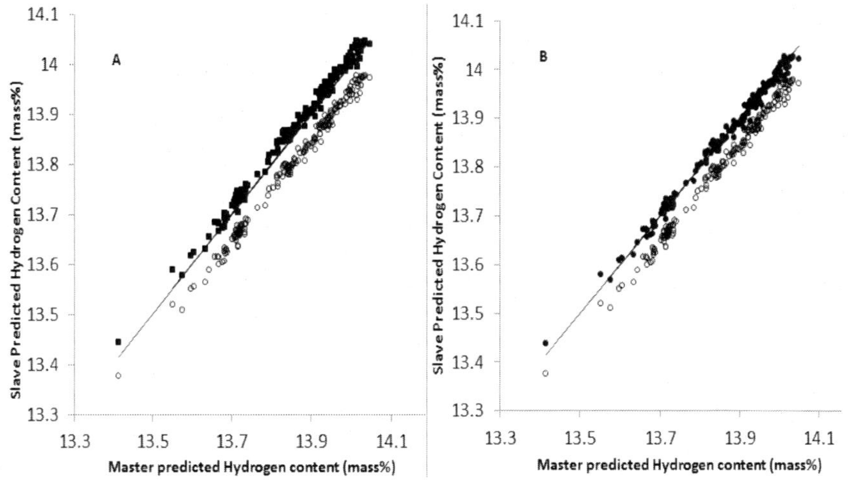

Figure 6. % Hydrogen Content predictions of test set fuels acquired on the secondary instrument for uncorrected (open circles), SVSSB corrected (solid squares) and BS corrected (solid circles) data. (A) Shows SVSSB correction and (B) shows SB correction using actual fuel standards. The solid line is an ideal line (slope=1 and intercept =0).

The effect of SVSSB as well as a simple slope bias (SB) correction using actual fuel standards is graphically depicted in Figures 4-6 for the API, cetane index, and % hydrogen content respectively. In all figures, “A” displays the SVSSB correction and “B” displays the SB correction using actual fuel standards. In addition, the open markers represent the uncorrected secondary instrument prediction and the solid marker represents the corrected secondary instrument correction. Also, for all Figures, the solid line represents an ideal relation between the primary and secondary predictions (slope =1 and bias = 0). As can be seen, for all of the property models, the SVSSB correction results in predicted values which more closely match that of the primary instrument's values for all jet fuels properties being modeled and in all cases the results are comparable to that of SB using actual fuel standards.

7.4.3. SVSSB Calibration Transfer between Different NIR-instruments

NIR spectra of 57 JP-8 jet fuels (TARDEC, Warren, MI, USA) were acquired on both FUElex NIR dispersive and MPA FT-NIR instruments to an independent test set. Raw JP-8 fuel spectra of both FUELex and MPA are shown in Figure 7, "A" shows a MPA jet fuel raw spectrum and "B" shows a FUELex jet fuel raw spectrum. As shown in Figure 7, the two spectra have different ranges and resolution. The FUElex instrument was used as the primary instrument since PLS models of the 3 shown jet fuel properties were built on it, and MPA is the secondary instrument. MPA spectra were splined to span the same wavelength range as FUELex (880-1570nm). The spectra were then pretreated by Savisky 1st derivative followed by SNV. A pretreated FUELex spectrum is overlaid with a pretreated MPA spectrum in Figure 8A, the solid line is the FUELex spectrum and the dashed line is the MPA pretreated spectrum. The pretreated spectra were then predicted on the PLS models. The errors of prediction are tabulated in Table 3. SEP is the standard error of prediction of the primary FUELex instrument relative to the actual values. For API, cetane index, and hydrogen content the SEP values are 0.27, 0.26, and 0.016 respectively, and all of SEP for the primary instrument are acceptable and fall within ASTM reproducibility values. Root mean square errors of prediction for the secondary MPA instrument ($RMSEP_U$) relative to the master instrument are 5.53, 4.30, and 0.23 for API, cetane index, and hydrogen content respectively, and all of the error are much higher than the accepted ASTM reproducibility given in Table 1.

Table 3. Standard Error of Prediction of the properties in the test set samples relative to the master instrument prediction for uncorrected and SVSSB corrected

Jet fuel property	Samples Number	Secondary Instrument			
		RMSEPu	RMSEPc	SB	r2
API Gravity	57	5.53	0.27	0.18	0.99
Cetane Index	57	4.30	0.26	0.25	0.99
Hydr. Cont.	57	0.23	0.016	0.035	0.99

RMSEPu: Root Mean Square Error of Prediction of uncorrected secondary instrument relative to primary instrument prediction; RMSEPc: Root Mean Square Error of Prediction of SVSSB corrected secondary instrument relative to primary instrument prediction; r2: correlation coefficient of primary prediction relative to secondary prediction.

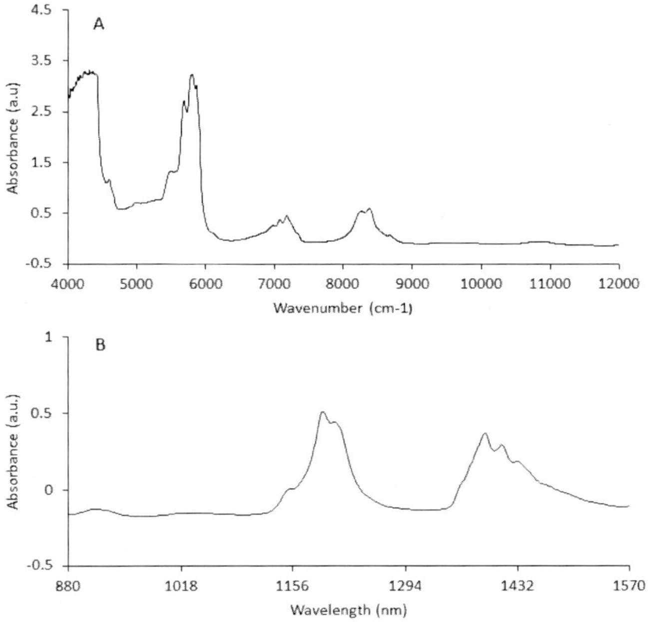

Figure 7. (A) gives NIR absorbance spectrum of a JP-8 fuel acquired on MPA FT-NIR instrument. (B) gives NIR absorbance spectrum of a JP-8 fuel acquired on FUELex dispersive NIR-instrument.

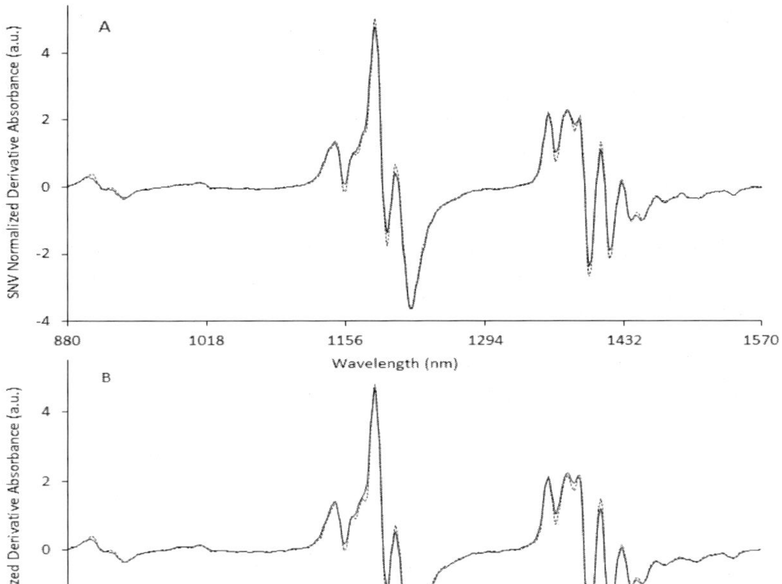

Figure 8. NIR absorbance spectra of a digitally synthesized MPA secondary instrument virtual standard (dotted line) overlaid with a digitally synthesized FUELex virtual standard (solid line) are given in (B) and their target fuel spectra of MPA (dotted line) and FUELex (solid line) are given in (A). Seven point Savitsky-Golay 1st derivative and SNV spectral transforms have been applied.

To apply SVSSB calibration transfer, the 15 neat chemicals were acquired on both FUELex and MPA instruments. The spectra of the chemicals were then pretreated and used to generated virtual standards as described previously. In Figure 8B, a virtual standard of the FUElex (solid line) is overlaid with a MPA virtual standard (dashed line). As can be seen in the Figure the difference between the two virtual standards (Figure 8B) is the same as real fuel spectra (Figure 8A). SVSSB calibration transfer was applied to the MPA instrument. Root mean square errors of prediction for the MPA secondary instrument after correction ($RMSEP_C$) is shown in Table 3 and are 0.27, 0.26, and 0.016 for API, cetane index, and hydrogen content, respectively. After applying SVSSB calibration transfer to the secondary instrument $RMSEP_C$ is now within the ASTM reproducibility. In addition, the calibration transfer was applied to the secondary instrument using the real fuel spectra and a simple slope bias correction (SB) to test the reliability of SVSSB. SB prediction errors are so close to SVSSB prediction errors except in the case of hydrogen content where the SVSSB prediction is better than SB.

The effect of SVSSB as well as a simple slope and bias using actual fuel standards (SB) correction is graphically depicted in Figures 9-11 for the API, cetane index, and % hydrogen content respectively. In all figures, “A” displays the SVSSB correction and “B” displays the SB correction. In addition, the open markers represent the uncorrected MPA secondary instrument prediction and the solid marker represents the corrected MPA secondary instrument prediction. Also, for all Figures, the solid line represents an ideal relation between the primary and secondary predictions (slope =1 and bias = 0). As can be seen, for all of the property models, the SVSSB correction results in predicted values which more closely match that of the primary instrument’s values for all jet fuels properties being modeled and in all cases the results are comparable to that of SB using actual fuel standards to transfer the calibration.

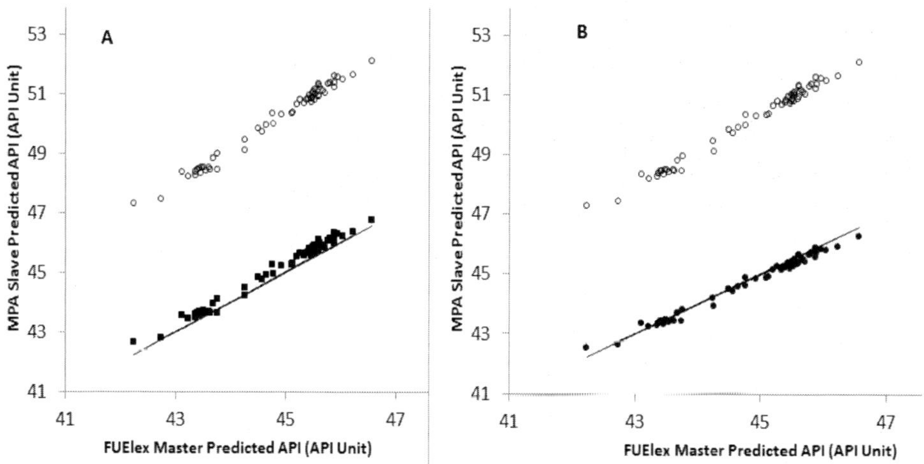

Figure 9. API predictions of test set fuels acquired on the MPA secondary instrument for uncorrected (open circles), SVSSB corrected (solid squares) and BS corrected (solid circles) data. (A) Shows SVSSB correction and (B) shows SB correction using actual fuel standards. The solid line is an ideal line (slope=1 and intercept =0).

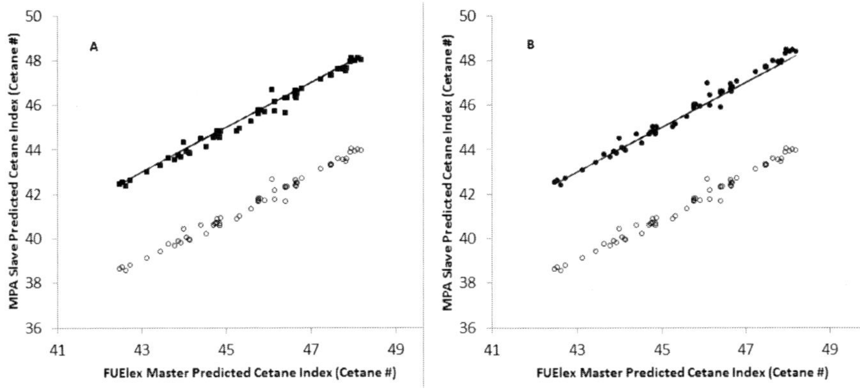

Figure 10. Cetane Index predictions of test set fuels acquired on the MPA secondary instrument for uncorrected (open circles), SVSSB corrected (solid squares) and BS corrected (solid circles) data. (A) Shows SVSSB correction and (B) shows SB correction using actual fuel standards. The solid line is an ideal line (slope=1 and intercept =0).

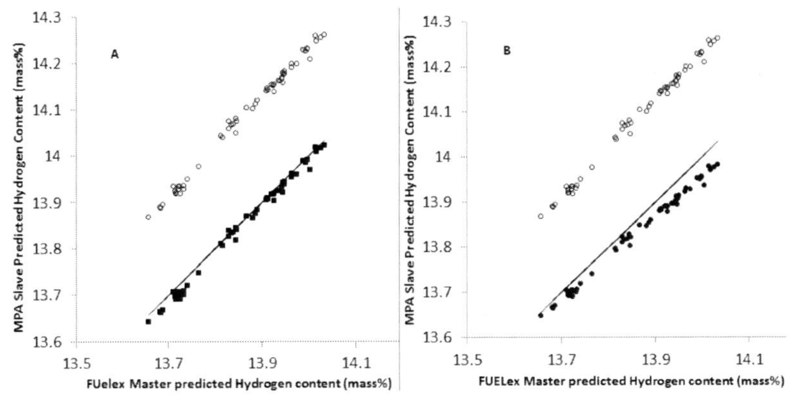

Figure 11. % Hydrogen Content predictions of test set fuels acquired on the MPA secondary instrument for uncorrected (open circles), SVSSB corrected (solid squares) and BS corrected (solid circles) data. (A) Shows SVSSB correction and (B) shows SB correction using actual fuel standards. The solid line is an ideal line (slope=1 and intercept =0).

CONCLUSION

Three PLS models of jet fuels properties were correctly transferred using virtual standards constructed using a segmented approach from 15 pure chemicals acquired on both the primary and a secondary instrument. SVSSB calibration transferred was carried out between two dispersive instruments and between dispersive and FT instruments. The RMS error of prediction of SVSSB corrected secondary instrument is similar to the RMS error of prediction for a slope and bias correction using real fuels. The use of the SVSSB approach eliminates the need to maintain fuel standards and a primary instrument, once the chemical spectra have been collected on the primary instrument. In addition, the acquisition of the spectra of 15 pure chemicals is all that is required to enable future calibration transfers to secondary instruments.

REFERENCES

Abdelkader, M. F., Cooper, J. B. & Larkin, C. M. Calibration transfer of partial least squares jet fuel property models using a segmented virtual standards slope-bias correction method. *Chemometrics and Intelligent Laboratory Systems* 110, 64-73, doi:10.1016/j.chemolab.2011.09.014 (2012).

Andrew, A. & Fearn, T. Transfer by orthogonal projection: making near-infrared calibrations robust to between-instrument variation. *Chemometrics and Intelligent Laboratory Systems* 72, 51-56, doi:DOI: 10.1016/j.chemolab.2004.02.004 (2004).

Cooper, J. B., Larkin, C. M. & Abdelkader, M. F. Calibration transfer of NIR partial least squares property models of fuels using virtual standards. *Journal of Chemometrics* 25, 496-505, doi:10.1002/cem.1395 (2011).

Cooper, J. B., Larkin, C. M., Schmitigal, J., Morris, R. E. & Abdelkader, M. F. Rapid Analysis of Jet Fuel Using a Handheld Near-Infrared (NIR) Analyzer. *Appl. Spectrosc.* 65, 187-192 (2011).

Cooper, J. B., Wise, K. L., Welch, W. T., Bledsoe, R. R. & Sumner, M. B. Determination of Weight Percent Oxygen in Commercial Gasoline: A Comparison between FT-Raman, FT-IR, and Dispersive NIR Spectroscopies. *Appl. Spectrosc.* 50, 917-921 (1996).

Cooper, J. B. *et al.* Comparison of NIR, Raman, and Mid-IR Spectroscopies for the Determination of BTEX in Petroleum Fuels. *Appl. Spectrosc.* 51, 1613-1620 (1997).

Cramer, J. A., Morris, R. E., Giordano, B. & Rose-Pehrsson, S. L. Partial Least-Squares Predictions of Nonpetroleum-Derived Fuel Content and Resultant Properties When Blended with Petroleum-Derived Fuels. *Energy & Fuels* 23, 894-902, doi:10.1021/ef800945c (2009).

Fernández-Ahumada, E., Garrido-Varo, A., Guerrero, J. E., Pérez-Marín, D. & Fearn, T. Taking NIR Calibrations of Feed Compounds from the Laboratory to the Process: Calibration Transfer between Predispersive and Postdispersive Instruments. *Journal of Agricultural and Food Chemistry* 56, 10135-10141, doi:10.1021/jf801881n (2008).

Pierna, J. A. F., Vermeulen, P., Lecler, B., Baeten, V. & Dardenne, P. Calibration Transfer from Dispersive Instruments to Handheld Spectrometers. *Appl. Spectrosc.* 64, 644-648 (2010).

Wang, Y., Veltkamp, D. J. & Kowalski, B. R. Multivariate instrument standardization. *Analytical Chemistry* 63, 2750-2756, doi:10.1021/ ac00023a016 (1991).

Bouveresse, E. & Massart, D. L. Improvement of the piecewise direct standardisation procedure for the transfer of NIR spectra for multivariate calibration. *Chemometrics and Intelligent Laboratory Systems* 32, 201-213, doi:Doi: 10.1016/0169-7439(95)00074-7 (1996).

Yee, N. G. & Coghill, G. G. Factor selection strategies for orthogonal signal correction applied to calibration of near-infrared spectra. *Chemometrics and Intelligent Laboratory Systems* 67, 145-156, doi:Doi: 10.1016/s0169-7439(03)00066-2 (2003).

Bouveresse, E. & Massart, D. L. Standardisation of near-infrared spectrometric instruments: A review. *Vibrational Spectroscopy* 11, 3-15, doi:Doi: 10.1016/0924-2031(95)00055-0 (1996).

Digital Filtering and Model Updating Methods for Improving the Robustness of Near-Infrared Multivariate Calibrations. Applied Spectroscopy 63, 246-255, doi:10.1366/000370209787392076 (2009).

Kunz, M. R., Ottaway, J., Kalivas, J. H. & Andries, E. Impact of standardization sample design on Tikhonov regularization variants for spectroscopic calibration maintenance and transfer. Journal of Chemometrics 24, 218-229, doi:10.1002/cem.1302 (2010).

Savitzky, A. & Golay, M. J. E. Smoothing and Differentiation of Data by Simplified Least Squares Procedures. Analytical Chemistry 36, 1627-1639, doi:10.1021/ ac60214a047 (1964).

Stewart, M. G. K. a. In Advanced Theory of Statistics. Charles Griffin and Company limited london 1 (1977).

In: Calibration Technology, Theories and Applications ISBN: 978-1-62618-808-2
Editors: Ikumatsu Fujimoto and Kunitoshi Nishimura

Chapter 8

CALIBRATION IN CHEMISTRY: STATE OF THE ART AND APPLICATIONS

Werickson Fortunato de Carvalho Rocha[1], Raquel Nogueira[2], Kássio Michell Gomes de Lima[3], Rodrigo Caciano de Sena[1], Márcia Silva da Rocha[1] and Joyce Rodrigues de Araújo[4]

[1]National Institute of Metrology, Quality and Technology (INMETRO), Directorate of Industrial and Scientific Metrology, Division of Chemical Metrology, Rio de Janeiro, RJ, Brazil

[2]National Institute of Metrology, Quality and Technology (INMETRO), General Coordination for Accreditation (CGCRE), Laboratory Accreditation Division (Dicla), Rio de Janeiro, RJ, Brazil

[3]Universidade Federal do Rio Grande do Norte, Instituto de Química, Grupo de Pesquisa em Quimiometria Aplicada, Natal, RN, Brazil

[4]National Institute of Metrology, Quality and Technology (INMETRO), Directorate of Industrial and Scientific Metrology, Division of Materials Metrology, Rio de Janeiro, RJ, Brazil

ABSTRACT

There is a need for improved quantitative information in science and technology. An increasing quantity of data has been generated in several levels of complexity, due to instrumentation advances in chemistry and chemistry-related fields. This has required and enabled the development of new procedures to explore analytical data in order to determine the properties of interest inside relatively complex systems. Calibration is an operation that, under given conditions, relates an output quantity to an input quantity for a measuring system. In chemical measurement processes, input quantities are given by analytical quantities that characterise certain types of analytes (chemical species), and their amounts or concentrations. Output quantities are represented by measured values, i.e. certain signals in positions z_j with intensities y_{zj}. The calibration methods can be divided regarding data complexity or dimensionality, in calibrations of zero, first and second orders. Zero order calibration is the univariate calibration method represented by for example a monochromatic photometer and a single constituent to be quantified. The classical statistical calibration procedure is to fit a linear regression of the instrument response (absorbance, in this case) on concentration, according to the Lambert-Beer law, and invert this relationship to determine the concentration in a new specimen. On the other hand, multivariate calibration constructs a predictive model on the basis of multiple spectral values. The multivariate calibration is divided in first and second calibration orders. Vector data for each sample belong to the first-order type, as for instance in the absorption or emission spectra (e.g., UV-visible spectrophotometry, spectrofluorimetry, infrared spectrophotometry). Matrix data for each sample belong to the second-order type, for instance, when two first-order instruments are coupled in tandem (e.g., hyphenated instruments such as LC-MS, GC-MS and GC-FTIR) and two-way arrays of data are generated. Second order calibration can also be produced using a single instrument, for instance a spectrofluorometer registering excitation-emission matrices (EEMs) or a diode-array spectrophotometer where a chemical reaction takes place.

The aim of this chapter is to review the key concepts of zero, first and second order calibration methods, based on two reference publications from the International Union of Pure and Applied Chemistry (IUPAC): K. Danzer and L. A. Currie (*Pure Appl. Chem.*, 1998, 70(4), 993–1014) and K. Danzer, M. Otto and L. A. Currie (*Pure Appl. Chem.*, 2004, 76(6), 1215–1225), and to demonstrate the application of these concepts by means of practical examples in pharmaceutical and inorganic analysis and polymeric materials. The chapter was divided into four sections: Section 1 discusses the theory of the different univariate and multivariate calibration methods applied to chemistry-related fields; Section 2 presents the application of different calibration methods in pharmaceutical and inorganic analysis for determination of active pharmaceutical ingredients (APIs), impurities, metals and metalloids in different types of matrices (e.g. food, water, beverages, fuels, minerals, drugs, pharmaceuticals products, clinical and biological materials); Section 3 discusses the calibration methods applied to polymeric materials characterization, as well as to the in-line monitoring of extrusion process of polymers, simulation and validation of mechanical properties of polymers, quantification of heavy metals in polymers and analysis of polymer additives; Section 4 shows the conclusions and future perspectives of different calibration methods applied to chemistry and chemistry-related fields.

Keywords: Calibration Methods; pharmaceutical and inorganic analysis; polymeric materials characterization

8.1. Calibration Methods in Chemistry

According to the guidelines for calibration in analytical chemistry-Part 1 [1] calibration is an operation that relates an output quantity to an input quantity for a measuring system under given conditions. In chemistry, input quantities are given by analytical quantities characterizing certain types of analytes (chemical species) q, and their amounts or concentrations x. Output quantities are represented by measured values, i.e., certain signals in positions z_j with intensities y_{zj}. The common case, in which a relationship has to be determined by calibration between the amount (content) of a certain analyte x_{qi}, and a signal intensity y_{zj} is only a special case of calibration. According to Escandar *et al.*[2] the various types of instrumental data have been classified on the basis of tensor algebra[3-5]. Within this scheme, when a given instrument produces a single instrumental response for a chemical sample, this datum is a scalar or zeroth-order tensor. Figure 1 shows an example of an absorbance of 0.8 at a wavelength of 250 nm showing the representation of a scalar. Vector data for each sample belong to the first-order type: for example, absorption (UV-visible or near-infrared spectroscopy). In the UV-visible or near-infrared (NIR) spectroscopy, the samples are exposed to radiation at different wavelengths (e.g., UV, Vis or NIR) and then generate different absorption spectra that can be used to identify or quantify sample components.

Figure 1 illustrates an example of UV-VIS spectrum in multiple wavelength showing the representation of a vector of data. When two first-order instruments are coupled in tandem (e.g., GC-MS, GC-MS-MS, LC-MS, LC-MS-MS), the order increases from first- to second-order. GC-MS, GC-MS-MS, LC-MS and LC-MS-MS are hyphenated techniques in which the sample components are first separated by means of gas chromatography (GC) or liquid chromatography (LC) and then detected by mass spectrometry (MS or MS-MS), with the objective to identify or quantify sample components. Second order calibration can also be produced using a single instrument: examples are a spectrofluorometer registering excitation-emission matrices (EEMs) or a diode-array spectrophotometer where a chemical reaction takes place.

Figure 1 shows an example excitation-emission matrix obtained by a spectrophotometer showing the representation of a matrix of data.

Grafical example	Data	Representation of data	Advantage	Spectroscopic data example
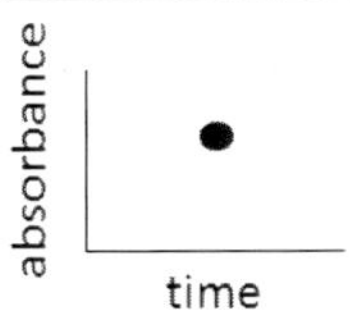	0th order	0.8 Scalar	————	UV abs. at 250nm
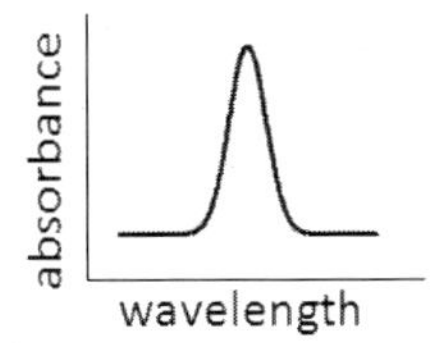	1st order	0.8 0.6 0.3 0.5 0.7 Vector	Outlier detection, interference compensation, noise reduction	UV spectrum
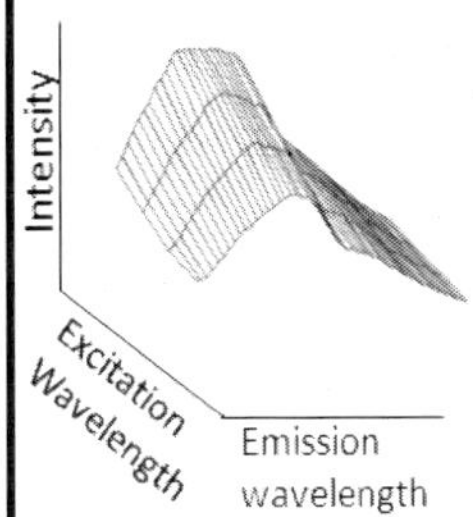	2nd order	0.8 1.1 0.7 0.6 1.3 0.5 0.3 1.0 0.6 ..... 0.7 0.9 0.5 Matrix	All of the above Plus unique solution	Fluorescence excitation-emission landscape

Figure 1. Different orders of instrumental data.

8.1.1. Zeroth-order Data

Zeroth-order data are usually fitted to a straight line by least squares, a procedure known as univariate calibration [1], which requires full selectivity for the analyte of interest. In the event that other sources respond, this signal should be constant, in order to be eliminated by a background correction or modeled by an intercept. When this is not the case, i.e., the responses are intrinsically unselective, a different approach, either instrumental or mathematical, is required.

When a calibration curve is a straight-line, we represent it using the following mathematical Equation 1:

$$y = ax + b \tag{1}$$

where y is the signal and x is the analyte's concentration. The constants b and a are, respectively, the calibration curve's expected y-intercept and its expected slope. The goal of a linear regression analysis is to determine the best estimates for b and a.

The most common approach to completing a linear regression for Equation 1 makes three assumptions:

(i) any difference between our experimental data and the calculated regression line is the result of indeterminate errors affecting y;
(ii) indeterminate errors affecting y are normally distributed;
(iii) the indeterminate errors in y are independent of the value of x.

Because we assume that the indeterminate errors are the same for all standards (homoscedasticity), each one contributes equally in estimating the slope and the y-intercept. For this reason, the result is considered an unweighted linear regression.

Homoscedasticity is one of the assumptions required in a regression analysis in order to make valid statistical inferences about population relationships. Homoscedasticity requires that the standard deviation and variance of the error terms (μ) are constant for all x, and that the error terms are drawn from the same population. This indicates that there is an uniform scatter or dispersion of data points about the regression line. According to Lavagnini and Magno [6] standard procedures can be followed to test whether homoscedasticity or heteroscedasticity holds:

(i) the plot of the residuals of the unweighted least squares regression versus the predicted values: a funnel shape trend with the increasing responses indicates an increasing variability with the concentration. More immediately the plot of the differences among the replicates and their mean at each concentration furnishes analogous results;
(ii) the Bartlett's test, which compares the variances of the replicates at each concentrations or a F-test between the largest and the smallest variance of the replicates [7]. In this context the Hartley's F_{max}-test could be also mentioned [8]. Under heteroscedastic conditions the plot of the experimental variances versus concentration and the calculation of the relevant model give some opportunities: the raw variance values coming from few repeated measurements are smoothed; the

availability of the relationship between the variance values and the concentration level makes easier the calculation of the weighted least squares regression and of its inverse as the weighting factors are the inverse of the variances at each concentration level.

8.1.2. First-order Data

First-order data belong to the multivariate domain, and are processed by suitable first-order multivariate calibration procedures. The central idea of all of these methodologies is to compensate for the lack of total selectivity of the analyzed signals, using efficient mathematical algorithms to extract the portion of the overall signal which can be employed to predict the concentration of a particular sample component. The most used first order multivariate calibration models are principal component regression (PCR) and partial least squares regression (PLS).

8.1.2.1. PLS

PLS regression is a linear multivariate data analytical method developed to handle data with high correlation, such as a NIR spectrum. In PLS a suitable set of latent variables are obtained using both independent (***X*** matrix) and dependent variables (***y*** vector) blocks by means of an iterative process that maximizes the covariance between these two blocks. The data matrix ***X*** is formed by the multivariate matrix (e.g. NIR spectra) and the vector ***y*** contains the reference values (e.g. concentration). Two sets of models corresponding to the linear algebraic relation between their scores are obtained of the form [9-13],

$$\boldsymbol{X} = \boldsymbol{T}\boldsymbol{P}^T + \boldsymbol{E} = \sum_{i=1}^{A} \boldsymbol{t}_i \boldsymbol{p}_i^T + \boldsymbol{E} \tag{2}$$

$$\boldsymbol{y} = \boldsymbol{T}\boldsymbol{q}^T + \boldsymbol{f} = \sum_{i=1}^{A} t_i \boldsymbol{q}_i^T + \boldsymbol{f} \tag{3}$$

where ***E*** (n x p) and ***f*** (n x 1) are error matrixes containing the parts of ***X*** (n x p) and ***y*** (n x 1), respectively, which are not explained by the model, n and p are the number of samples (rows) and variables (columns), respectively. $\boldsymbol{t}_i$ is the column vector that comprise the score matrix ***T*** (n x A). p_i and q_i are the loading that comprise ***P*** (p x A) and ***q*** (1 x A) loading matrixes, respectively, with A equal to the number of latent variables suitable to explain the variance of the variables. The Figure 2 represents these matrixes.

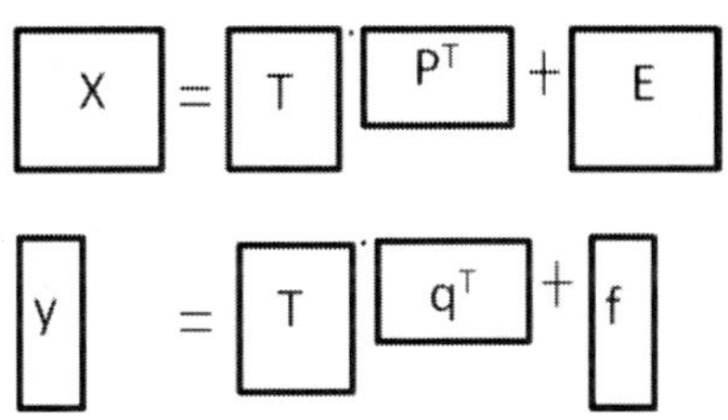

Figure 2. Decomposition of matrix *X* and vector *y* on their scores and loadings.

The parameters of interest ($\hat{y}$), for a set of samples, are obtained by multiplication of the *X* matrix by an suitable regression vector *b* (p x 1), calculated as:

$$\hat{\mathbf{y}} = \mathbf{Tq}^{\mathrm{T}} = \boldsymbol{XW}(\boldsymbol{P}^T\boldsymbol{W})^{-1}\boldsymbol{q}^T = \boldsymbol{Xb} \quad (4)$$

Where *W* (p x A) is the weight matrix obtained by the PLS algorithm.

8.1.2.2. PCR

The purpose of principal component regression (PCR) is to estimate the values of a response variable at the basis of selected principal components (PCs) of the explanatory variables. There are two main reasons for regressing the response variable on the PCs rather than directly on the explanatory variables. Firstly, the explanatory variables are often highly correlated (multicollinearity) which may cause inaccurate estimations of the least squares (LS) regression coefficients. This can be avoided by using the PCs in place of the original variables since the PCs are uncorrelated. Secondly, the dimensionality of the regressors is reduced by taking only a subset of PCs for prediction.

8.1.3. Second-order Data

Second-order data, when there is one matrix per sample (see Figure 1). When second-order data for a group of samples are joined, they can be arranged maintain in grows or columns in common, or in a three-dimensional array. The result is known as three-way array. This approach can be seen in the Figure 3. Several multivariate calibration models to data of second order is being used, such as: Parallel Factor Analysis (PARAFAC), Tucker3, Regression Partial Least Squares N-dimensional (N-PLS) and Bilinear least Squares (BLLS).

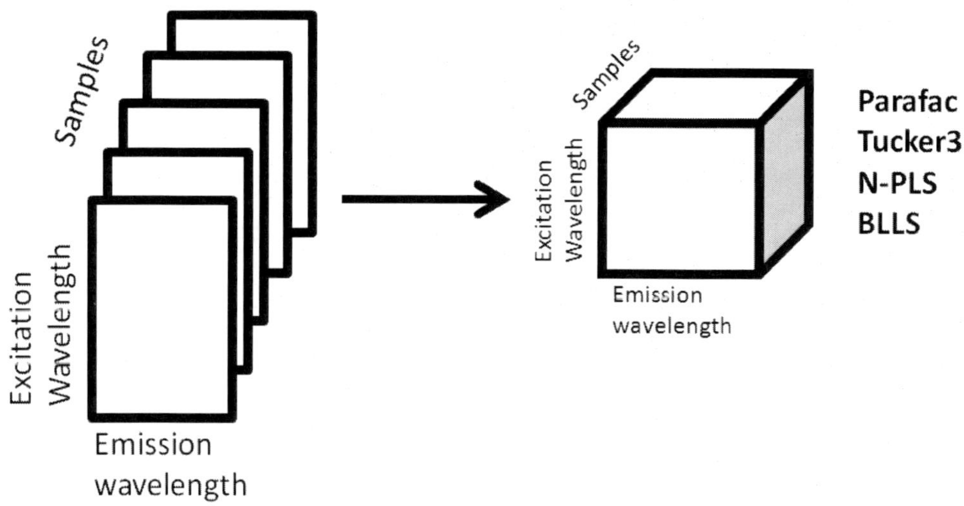

Figure 3. Arrangements of second-order data.

8.1.3.1. Parafac

Second-order data are obtained when a given sample produces a *J* x *K* data matrix or second-order array, where *J* and *K* denote the number of data points in the first and second dimensions, respectively (in EEM fluorescence measurements, *J* is the number of digitized emission wavelengths and *K* is the number of excitation wavelengths). If the *I* training matrices and the unknown sample matrix are stacked, a three-way data array *X* is obtained, whose dimensions are [(*I* + 1) x *J* x *K*]. Provided *X* follows the trilinear PARAFAC model [14], it can be written as a sum of Kronecker products of three vectors for each responsive component: a_n, b_n, and c_n, which collect the relative concentrations or scores [(*I* + 1) x 1], the emission profiles (*J* x 1), and the excitation profiles (*K* x 1) for component *n*, respectively. The specific expression is thus:

$$X = \sum_{n=1}^{N} a_n \otimes b_n \otimes c_n + \underline{\mathbf{E}} \tag{5}$$

where ⊗ indicate the well-known Kronecker product, *N* is the total number of responsive components, and $\underline{\boldsymbol{E}}$ is a residual error term of the same dimensions as ***X***. The column vectors a_n, b_n, and c_n are usually collected into the loading matrices *A*, *B* and *C*, respectively.

Figure 4 shows the graphical representation of Equation 5 indicating the decomposition of a three-dimensional data array F triads vector weights. Each triadis equivalent to a factor PARAFAC model. Alternatively, the PARAFAC model can be represented by the following matrix equation:

$$\underline{\mathbf{X}} = \mathbf{A}(\mathbf{C}|\otimes|\mathbf{B})^t + \underline{\mathbf{E}} \tag{6}$$

where matrices A, *B* and C have dimensions (*I* x *F*), (*J* x *F*), (*K* x *F*), respectively, and the symbol "|⊗|" represents the product of Khatri-Rao.

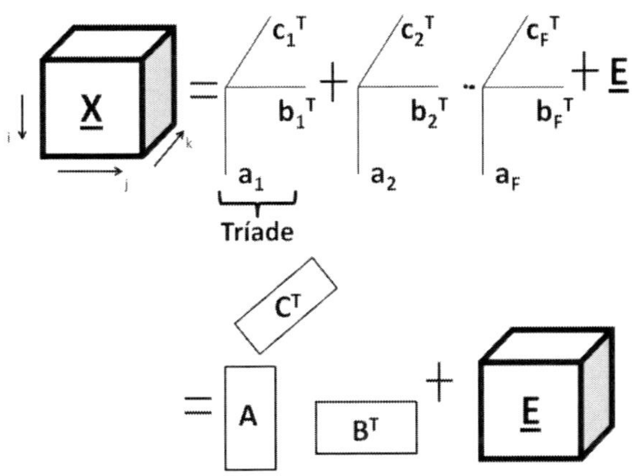

Figure 4. Graphical representation of PARAFAC model. Decomposing three-dimensional data array F triads of weight vectors.

PARAFAC is the most frequently used multiway technique in analyzing three-way excitation emission data because of the trilinear nature of most of these data sets, which is a necessary condition to obtain reliable results from this method. PARAFAC analysis of trilinear excitation–emission data sets, with or without a low level spectral overlap between no interacting fluorophores, ensures unique and acceptable results. However, there may be some situations in which all of these conditions are not fulfilled, such as relative concentration profiles of the fluorophores or FRET between a donor–acceptor fluorophorescouple [15-18]. In such cases, depending on the degree of deviation from the trilinearity, PARAFAC may fail to resolve the data.

8.1.3.2. Tucker

Tucker3 method can be used for compression and data exploration of N-way array [19,20]. The mathematical notation of Tucker3 model is illustrated in Equation 7:

$$\underline{X} = AG(C|\otimes|B)^{T} + \underline{E} \tag{7}$$

where $\underline{\boldsymbol{X}}$ is the original data array, G is a core array with dimensions of chemical ranks of modes, A, B and C are the loading matrices of the first, second and the third modes, respectively, and $\underline{\boldsymbol{E}}$ is the residual matrix. The core array encompasses the inherent interactions of different loadings and approximates the variation of $\underline{\boldsymbol{X}}$.

The three-way Tucker model is a generalization of the trilinear PARAFAC model. A data cube, R, is decomposed into three sets of factors, $\hat{x}$, $\hat{y}$, and $\hat{z}$, as with PARAFAC. However, the Tucker3 model differs from the PARAFAC model in two key ways. The number of factors in each order of the Tucker3 model is not constrained to be equal. Also, the Tucker3 model employs a small core cube, C, that governs the interactions among the factors. A nonzero element in the p_{th}, q_{th}, and r_{th} positions of the core C dictates an interaction between the p_{th} factor in the Xway, the q_{th} factor in the Y-way, and the r_{th} factor in the Z-way.

If the core matrix is unconstrained, that is, the core is free to mix the X, Y, and Z modes in any manner necessary to fit the data in a least squares sense, the resolved factors are rotationally ambiguous just as are the factors determined by principal component analysis (PCA). However, when elements of C begin to be constrained, the rotational ambiguity is confined up to the point that the Tucker3 model becomes unique to within scaling factors of $\hat{\mathbf{X}}$, $\hat{\mathbf{Y}}$, and $\hat{\mathbf{Z}}$. If the core matrix is constrained to be all 0 except for the superdiagonal elements, p=q=r, the Tucker model reduces to the PARAFAC model. Constraining the Tucker model permits analysis of chemical data where two or more factors might have the same chromatographic profile but different spectral and concentration profiles. In the instances where two or more chemical components share a pure profile in one order, the data become rank deficient with respect to the trilinear model. In many cases, the constrained Tucker3 model retains the second-order advantage associated with trilinear data.

Despite the fact that Tucker3 method can handle nontrilinear data sets, it has gained less attention because of some of its shortcomings such as rotational freedom that causes nonunique results and also complications in interpretation of the obtained models arising from a high number of interaction terms between the factors. Considering the fact that PARAFAC is too restricted for analyzing many complicated data sets and Tucker3 is too flexible, a restricted Tucker3 model was proposed as a hybrid method to gather some of the advantages of both methods in analyzing data sets with deviation from trilinearity [21].

8.1.3.3. N-PLS

N-PLS is an extension of the bilinear PLS algorithm to be applied for multi-way array [14]. The principles are similar to the PLS algorithm, such that N-PLS also uses dependent and independent variables for finding latent variables that describe their pair-wise maximal covariance, and is a sequential algorithm thus the F-1component model is a subset of the F component model. N-PLS decomposition starts by constructing a distinct PARAFAC like model for the dependent response variables (***Y***) and for the independent descriptor variables (***X***) and maximizing the covariance between these two matrices. The mathematical notation of the N-PLS calibration model can be written as:

$$\underline{X} = T(W^k|\otimes|W^J)^T + \underline{E_X} \quad (8)$$

$$\underline{Y} = U(Q^M|\otimes|Q^L)^T + \underline{E_Y} \quad (9)$$

$$U = TB + E_U \quad (10)$$

where matrices ***T*** and ***U*** include score vectors of the original data, ***W*** and ***Q*** weight vectors, $\underline{\boldsymbol{E}}_i$ residual terms and *B* regression coefficients.

8.1.3.4. BLLS

In contrast to PARAFAC, concentration information is introduced into the BLLS [22,23] calibration step (without including data for the unknown sample), to obtain approximations to pure analyte matrices at unit concentration. For this purpose, the concentration product matrix ***D*** and N_{cal} concentration-weighted $\boldsymbol{T_n}$ matrices are first obtained (N_{cal} is the number of calibrated analytes), employing data from the ***I*** training matrices $X_{i,cal}$ and the calibrated analyte concentrations contained in the $I \, \mathrm{x}_{cal}$ matrix ***Y***:

$$T_n = \sum_{i=1}^{I} Y_{in} X_{i,cal} \quad (11)$$

$$D = Y^T Y \quad (12)$$

The required $\boldsymbol{S_n}$ matrices for analyte *n* at unit concentration are then obtained as:

$$Sn = \sum_{n'=1}^{N} (D^{-1})nn' T_{n'} \quad (13)$$

Notice that in the presently studied analytical problem, a single calibrated analyte occurs. Hence, $N_{cal} = 1$, $n = 1$, and T_n should be replaced by T_1, Y by the vector $\boldsymbol{y}$ of known standard concentrations, as used above for PARAFAC, *D* by the scalar quantity *D*, and S_n by S_1:

$$T_1 = \sum_{i=1}^{I} y_i X_{i,cal} \quad (14)$$

$$D = y^T y \quad (15)$$

$$S_1 = \sum_{n=1}^{N} D^{-1} T_1 \quad (16)$$

8.2. Application of Different Calibration Methods in Pharmaceutical and Inorganic Analysis

8.2.1. Calibration in Pharmaceutical Analysis

The quality control of pharmaceuticals is primarily based on pharmacopeial requirements for active pharmaceutical ingredients (APIs), excipients and finished products (also applicable to bulk products). Quality control laboratory analysis consists mainly of tests to determine the API mass fraction (% m/m), to identify (and, if possible, quantify) impurities, as well as to check physico-chemical and microbiological properties.

Several techniques are used to quantify APIs and impurities. The main impurities of interest are the so-called related substances, which comprise API degradation products and impurities originated from the API synthetic pathway, as well as residual solvents.

The commonly used calibration approaches are discussed below.

8.2.1.1. Single-point Calibration in Pharmaceutical Analysis

Single-point calibration assays are the common pharmacopeial approach used to calculate API mass fractions. In this method, analyte and reference solutions are prepared in a single concentration level and then the analyte concentration is estimated from the analyte and standard responses, standard solution concentration and standard purity. Pharmacopeial assay methods are usually accompanied by equations that allow the estimation of the API mass fraction (or purity), such as Equation 17:

$$w_{API} = (R_{API} \text{ x } w_S \text{ x } P_S) / R_S \quad (17)$$

where w_{API} is the API mass fraction, R_{API} is the analytical response for the API solution, w_S is the reference standard solution mass fraction, P_S is the reference standard purity and R_S is the analytical response for the reference standard solution.

8.2.1.2. Univariate Calibration in Pharmaceutical Analysis

In the univariate calibration, the concentration of a single compound can be determined by using, for instance, the spectrometer response or the chromatographic peak area at a certain wavelength [12]. If several external standard solutions are analysed (at least three, but usually five), a calibration curve may be constructed ($y = ax + b$). After confirming the adequacy of the calibration curve based on its correlation coefficient and absence of tendencies, as well as definition of the quantitation limit (LOQ) and working range, the line is then used to determine the API concentration in samples. The best estimate for the intercept *b* is obtained when the best fit straight line is not forced to pass through the origin [12].

It should be noted that, even if linear calibration curves are predominant, the use of quadratic or even other relationships is possible [24]. In the case of high-performance liquid chromatography combined with an evaporative light scattering detector (HPLC-ELSD), non-linear calibration curves are mainly the rule than the exception.

In the classical univariate calibration, which is the traditional method used in analytical chemistry books, the error refers to the y-axis, which usually expresses responses dependent on equipment performance. On the other hand, chemometrists usually employ inverse univariate calibration. In this case, errors are in the x-axis, which expresses API concentration or mass fraction and to which the most important sources of error are related [12].

The use of derivatives is an important mathematical feature traditionally applied to analytical measurement results. By using first or higher derivatives, it is possible to improve method selectivity, since this mathematical treatment has an enhanced ability to detect and to measure minor spectral features, making possible the distinction between quite similar spectra and the quantitative analysis even in the presence of overlapping responses [25,26]. However, it should be considered that random measurement errors are aggravated in the derivative mode [25].

Univariate calibration is also carried out by the standard-addition method, which consists of adding known and increasing standard amounts to fixed sample solution volumes, followed by dilution to make up a fixed final volume. Usually, the standard is the pure form of the analyte. The calibration curve "peak area" *versus* "added standard concentration" moves upwards in the graph, in case the analyte under study is present in the sample, and the analyte concentration corresponds to the curve intercept *b* with the y-axis. One important application of such a method is to neutralise possible "matrix effects", since both the analyte and standard solutions contain similar matrix amounts [27].

It is also possible to add a known and constant amount of an internal standard (IS) to sample solutions, in order to correct for any analyte loss during sample preparation or differences between equipment runs. The IS is usually similar but not identical to the analyte under study and has to provide a response that can be distinguished from that of the analyte. The calibration curves, in this case, are constructed by plotting the relative peak area ratio analyte/IS as a function of analyte concentration.

8.2.1.3. Multivariate Calibration in Pharmaceutical Analysis

Unlike univariate calibration methods, which require "an interference free" system, multivariate calibration can deal with results from samples containing more than one API, interfering excipients or impurities, even if these responses are strongly overlapped [28], and also with responses measured on multiple wavelengths [29]. Therefore, multivariate calibration allows the simultaneous analysis of mixture components without using chemical pre-treatments, derivatives, ratio spectra spectrophotometry or prior chromatographic separation [28]. The method consists of constructing a mathematical model that relates a property (e.g. API mass fraction) to analytical responses (e.g. absorbances) by using reference samples (calibration step), and then using the model to estimate the API concentration (prediction step) [30]. It should be noticed that the introduction of a new constituent in the sample being analysed makes recalibration necessary [27].

8.2.1.4. Applications

The usual pharmacopeial assay methods are based on titrimetry, spectrophotometry, high-performance liquid chromatography coupled with detectors such as UV/Vis, diode array (HPLC-DAD), fluorescence, HPLC-ELSD, refractive index (HPLC-RID) and gas chromatography (GC) with flame ionisation detector and headspace injection (headspace GC-FID) (quite useful for analysis of volatiles in non-volatile pharmaceuticals). Other analytical techniques, even if not recommended by most of the pharmacopeias, find large application, including liquid and GC coupled to mass spectrometry (LC-MS, LC-MS/MS, GC-MS, CC-MS/MS), capillary electrophoresis (CE), Raman spectroscopy and quantitative nuclear magnetic resonance (qNMR). In most of the pharmacopeias, infrared spectroscopy (IR) is primarily dedicated to qualitative tests (identity tests) and its potential for quantitative analysis, especially when combined with chemometric methods, has been neglected. The use of calibration methods applied to such techniques will be discussed in this chapter section.

Although the UV-Vis spectroscopy is a quite simple and cheap technique, it is quite often not an appropriate analytical procedure, due to the presence of substances that absorb at the same wavelength as the analyte, including not only pharmaceutical excipients, but also API impurities, which usually have quite similar structures (and chromophores) compared to the corresponding analytes. This is the main reason why UV-Vis spectroscopy is considered a non-selective method. If no interference between the spectra of the analyte and other sample components is observed, the assays are carried out by such usual procedures as single-point calibration or external calibration curves. Quite often, even-numbered derivatives (e.g. first derivatives) are also used for spectrometric data processing, as in the quantitative analysis of the antidepressant drugs nefazodone hydrochloride and sertraline hydrochloride [31] and in the determination of benzophenone, which is the main impurity in phenytoin bulk powder and pharmaceutical formulations [32]. Besides improving spectra resolution, even-numbered derivatives are also reported to reduce irrelevant absorbance due to excipients in pharmaceutical dosage forms [31].

Additionally, if UV-Vis spectroscopic measurements are combined with chemometrics (multivariate calibration methods), the simultaneous determination of two or more analytes in pharmaceutical preparations by UV-Vis spectroscopy, even in the presence of overlapping signals and without prior chromatographic separations (e.g. by HPLC), can be carried out. Some examples include:

(i) UV determination of nifuroxazide and drotaverine hydrochloride in formulations for treatment of spasmodic diarrhoea, even if these two APIs absorb in the same UV spectral region, by means of application of the multivariate calibration methods of partial least squares (PLS), classical least squares (CLS) and principal component regression (PCR) [33];

(ii) use of the multivariate regression methods of support vector regression (SVR) and artificial neural networks (ANNs) to determine mebeverine hydrochloride and sulpiride in binary mixtures in the presence of their impurities and degradation products with superimposed UV spectra [34]; and

(iii) analysis of thiamine and pyridoxine hydrochlorides using UV-Vis spectrophotometry combined with CLS and genetic algorithm-based multivariate calibration methods [29]. Multivariate calibration methods can also be applied to first-order derivative UV-spectra, as demonstrated in the simultaneous determination of phenytoin, barbital and caffeine [35], and also of chlorhexidine hydrochloride and lidocaine hydrochloride [36].

The usual procedure for API determination by high-performance liquid chromatography (HPLC) or ultra-high performance liquid chromatography (UHPLC) is separation using specific columns (reversed phase, normal phase and ion-exchangers, among others) and detection and construction of external calibration curves by linear regression analysis, from which the analyte concentration can be estimated. Figure 5 shows examples of HPLC-DAD chromatograms used for determination of three APIs and their main impurities, with their corresponding external calibration curve. Other examples given in the literature include determinations by RP-HPLC-UV of ketoconazole, in the presence of four related compounds and excipients [37], mosapride and pantoprazole in pharmaceutical tablets and plasma samples [38] and sibutramine and seven other substances in slimming products [39], as well as the determination by LC-MS/MS of gentamicin, lincomycin and spectinomycin in the presence of their impurities in pharmaceutical formulations [40]. In all cases, external standard calibration curves of peak areas as a function of concentration were constructed for each drug and then used to estimate the API concentrations.

In other RP-HPLC-UV analyses, standard and sample solutions were "spiked" with a fixed amount of an IS and the external calibration curves were constructed by plotting peak area ratios of the analyte/internal standard versus concentration. This procedure was applied to determine the two drugs used in the hypercholesterolemia treatment, rosuvastatin and atorvastatin, using naproxen sodium as the IS [41], to quantify the non-steroidal anti-inflammatory agent etoricoxib using bromhexine as the IS [42] and in the analysis of the antihistamine drug fexofenadine hydrochloride and its related compounds, after addition of the IS lisinopril [43].

If a diode array detector (DAD) is connected to the HPLC, it is possible to use multiple linear regression (MLR), in which two or more wavelength responses are used to estimate the API content. Some authors denominate chromatographic data from a single wavelength detection as first-order data, while second-order data is a term reserved for chromatographic data at multiple wavelengths [44]. The advantage of MLR is that the use of several wavelengths will often provide better results than relying on a single wavelength [12], in terms of sensitivity [44] and due to smaller errors compared to single regression equations based on a single wavelength [45]. An example was given in the analysis of naproxen sodium and pseudoephedrine hydrochloride in tablets by RP-HPLC-DAD, using losartan potassium as the IS, and in combination with CLS, PCR and PLS to allow the interpretation of multiple wavelengths [45]. By means of multivariate calibration methods, it is still possible to use the whole spectrum in each wavelength in the calculation [12].

The application of multivariate calibration methods to HPLC chromatographic results makes possible the simultaneous determination of two or more APIs, as in the analysis of enalapril maleate and nitrendipine in tablets by RP-HPLC-DAD combined with PLS regression, using dianogest as the IS [28]. It also enables the quantification of small impurity peaks that co-elute with the other peaks in HPLC-DAD chromatograms [12], as in the quantification of trace amounts of the anti-epileptic drug phenytoin by three-dimensional HPLC-DAD, in the presence of interferent peaks, with accuracy comparable to LC-MS/MS results [46]. The combination of HPLC-DAD separation and chemometrics was also successful to determine ibuprofen enantiomers in pharmaceutical formulations, after HPLC separation using a chiral chromatographic column and the use of PLS regression, even if the enantiomers' chromatographic profiles were strongly overlapped [44].

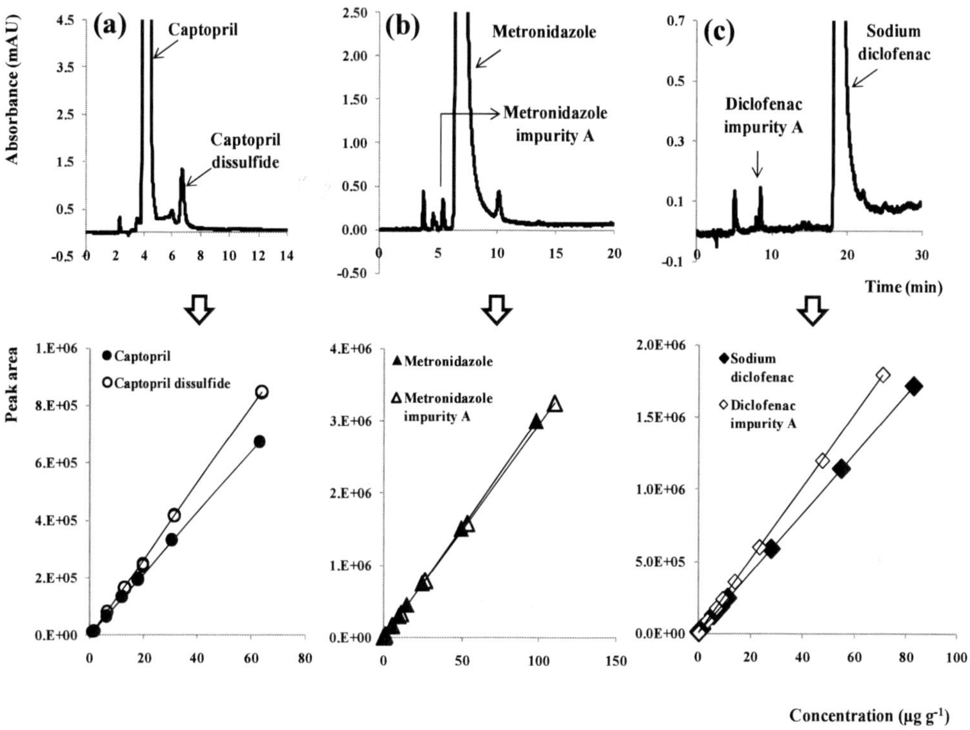

Figure 5. Expanded chromatograms and external calibration curves for HPLC-DAD determinations of captopril and its main impurity captopril disulfide (a), metronidazole and metronidazole impurity A (b) and sodium diclofenac and diclofenac related compound A (c). Experimental conditions (a) RP-18 column, 25 cm x 4.6 mmi.d., 5 μm; mobile phase, methanol - 0.11 % phosphoric acid (volumetric ratio 55:45); flow-rate, 1 $mLmin^{-1}$; detection wavelength, 220 nm; injection volume, 20 μL; (b) RP-18 column, 25 cm x 4.6 mmi.d., 5 μm; mobile phase, 1.36 g L^{-1} potassium dihydrogen phosphate - methanol (volumetric ratio 70:30); flow-rate, 1 $mLmin^{-1}$; detection wavelength, 315 nm; injection volume, 10 μL; (c) RP-18 column, 25 cm x 4.6 mmi.d., 5 μm; mobile phase, methanol - 0.005 $molL^{-1}$ potassium dihydrogen phosphate pH 2.5 (volumetric ratio 70:30); flow-rate, 1 $mLmin^{-1}$; detection wavelength, 254 nm; injection volume, 20 μL.

Concerning spectrofluorimetry, several references were found in the literature about univariate calibration, including the assay of the anti-hemorrhagic drug ethamsylate first by derivative synchronous fluorometry [47] and the determination of the antipsychotic and the antiemetic drug levosulpiride after complexation with the metal ions Fe^{+3} and Al^{+3}[48]. Other examples of univariate calibrations after RP-HPLC separation and spectrofluorimetric detection, using ISs, are the determination of methocarbamol and aspirin [49] and the analysis of the anti-malarial agents quinine and chloroquine, using salicylic acid as the IS [50]. The use of chemometrics for the treatment of fluorescence spectroscopy results is well exemplified by Fang et al. [51], who used parallel factor analysis (PARAFAC) and alternating trilinear decomposition algorithm (ATLD) for the quantification of ofloxacine, norfloxacine and enoxacine in human plasma. Darwish and Backeit [52] compared the application of four multivariate calibration methods, namely PLS, genetic algorithm and PLS (GA-PLS), PCA-ANN and GA-ANN, as well as of two classical univariate calibration methods (direct and isoabsorptive), to the spectrofluorometric analysis of olmesartanmedoxamil (an angiotensin II receptor blocker) and amlodipine besylate (a calcium channel blocker) in a combined dosage form for the treatment of hypertension, and reported the superiority of the multivariate calibration models over the univariate ones in terms of accuracy and precision.

Since most active pharmaceutical ingredients are non-volatile, thermally unstable or both, the use of CG is not usual for the determination of API purity in raw materials or API quantification in bulk or finished products. However, the literature contains numerous articles that deal with the quantification of pharmaceutical drugs in wastewater, which usually requires extraction and derivatisation procedures prior to GC analysis, as exemplified by Shin and Oh [53] for the determination of thirteen non-steroidal anti-inflammatory drugs (NSAIDs) in river water by GC-MS. However, GC finds important application in the pharmaceutical analysis of residual solvents, which can be carried out by static headspace GC-FID (sH-CG-FID) or dynamic headspace GC-FID (purge-and-trap). In the sH-CG-FID, the dissolved sample is heated in a closed vial until an equilibrium is established between the upper gas phase and the lower liquid phase, and then a fixed gas phase volume containing only volatile components is injected into the GC column, while in the purge-and-trap method, a gas flow is swept over the sample in order to release the volatiles, which are trapped in a specific column (e.g. Tenax^R) and then thermally desorbed to the gas spectrometer [54,55]. Both external and internal calibration methods can be used in headspace GC analysis [56,57]. Figure 6 shows external calibration curves for several solvents for analysis of residual solvents by headspace GC-FID in APIs.

Univariate and multivariate calibration methods are also applicable to assays by capillary electrophoresis (CE). The use of external standard calibration curves is a common procedure, as in the determination by CE of the hazardous impurities dimethyl sulfate and chloroacetyl chloride present in trace levels in drug substances [[58]].

Infrared spectroscopy (IR) is divided into three ranges: near infrared (12.821–4.000 cm^{-1} / 800–2500 nm), middle infrared (4.000–400 cm^{-1} / 2.500–25.000 nm) and far infrared (400–10 cm^{-1} / 100–1000 μm). Near-infrared spectroscopy (NIR) finds extensive application in the quality control of pharmaceutical raw materials [59,60], bulk [61,62] and finished products [63,64]. The great potential of NIR appears when it is combined with chemometrics, in order to allow the interpretation of broad and highly overlapping bands in the NIR spectrum range [65,66] and to enable collinearity correction [66]. The resulting technique is considered to be simple, rapid, non-invasive, non-destructive and to require no or little sample-pretreatment [66,68]. Due to these properties, NIR-chemometrics can be used as a process analytical technology (PAT) tool for real-time control of critical production steps, as proposed by the Food and Drug Administration (FDA) [67, 69, 70]. It also increases the potential for 100% inspection of pharmaceuticals [71]. However, chemometric methods are still not included in the pharmacopeias [68].

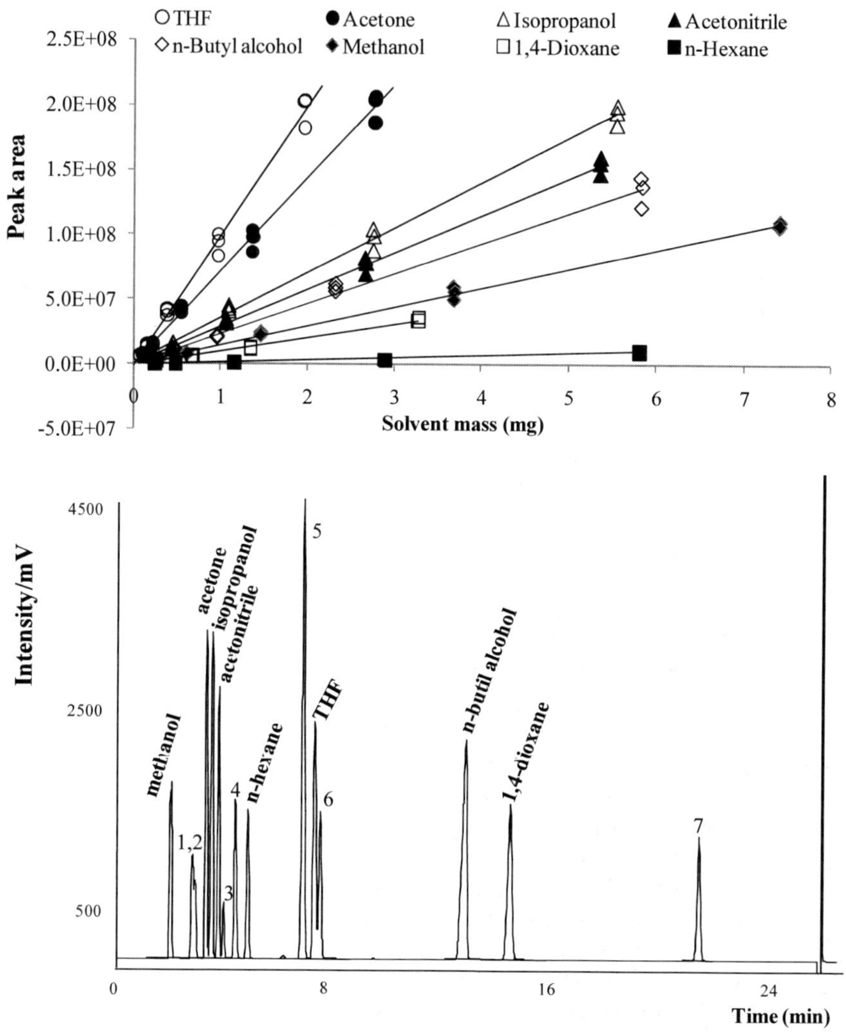

Figure 6. Headspace GC-FID external calibration curves for eight residual solvents and chromatogram indicating the elution position of these solvents: methanol (a) (1.67 μg g^{-1}), acetone (b) (0.56 μg g^{-1}), isopropanol (c) (1.11 μg g^{-1}), acetonitrile (d) (1.11 μg g^{-1}), *n*-hexane (e) (1.11 μg g^{-1}), THF (f) (0.28 μg g^{-1}), *n*-butyl alcohol (g) (1.67 μg g^{-1}) and 1,4-dioxane (h) (1.67 μg g^{-1}). The other solvents in the chromatogram are: ethanol (1) (0.56 μg g^{-1}), diethyl ether (co-eluted with ethanol) (2) (0.28 μg g^{-1}), dichloromethane (3) (1.67 μg g^{-1}), methyl *tert*-butyl ether (4) (0.56 μg g^{-1}), ethyl acetate (5) (0.56 μg g^{-1}), chloroform (6) (0.14 μg g^{-1}) and toluene (7) (0.83 μg g^{-1}). Experimental conditions: WCOT fused silica CP-select 624 (Varian) (6 % cyanopropyl phenyl and 94 % dimethyl polysiloxane), 30 m x 0.32 mm i.d., DF 1.8; helium gas-flow, 2.20 mL min^{-1} (35 cm s^{-1} at 40 °C); GC oven temperature gradient, 40° C (holdtime 20 min), 10° C min^{-1} until 240° C (hold time 20 min); injector temperature, 140 °C; FID temperature, 250° C; headspace thermostatting temperature, 80° C; headspace thermostatting time, 60 min; needle temperature, 85° C; pressurization time, 30 s; injection volume, 1 mL; spit-ratio 1:5. The solvent solutions were prepared in a mixture dimethylformamide-water (5:1, v/v).

A detailed description of chemometric methods applicable to NIR interpretation can be found at Roggo et al. [72and numerous examples of NIR-chemometrics applied to pharmaceutical analysis can be found in the literature, for instance in the analysis of neomycin [73], irbesartan, paracetamol [74], ibuprofen [75], teophylline [71], nimesulide [76] and amoxicillin [68]. Some of these methods were presented as alternatives to pharmacopeial assay methods, in substitution for instance of the labourious microbiological assay of neomycin [73]. Multivariate calibration models applied to near-infrared spectra were also used for non-invasive determination of moisture in freeze-dried mannitol-containing products [77].

At this point, it should be discussed that multivariate calibration methods, when applied to NIR and also to other analytical techniques, require special attention in certain aspects. When the analytical calibration model is constructed, it is essential to use a representative calibration set, which usually comprises a range of $\pm$ 20% around the nominal value and will ideally take into account the variability in the production samples [74]. In the case of tablets, significant differences were reported between the calibration carried out under one set of conditions (primary conditions) and the analysis of production tablets obtained under other (secondary) conditions, due to differences in equipment, environmental conditions and tablets' composition [78]. Second, it may be necessary to remake the calibration model on a regular basis, since it depends on equipment performance [12]. Finally, multivariate calibration may give only a rough prediction of the API content and dubious results may require more investigation using other analytical techniques, such as chromatography [12].

Raman spectroscopy, even if not described in the pharmacopeias, finds large application both for qualitative (identification) or quantitative purposes. An interesting example of quantitative analysis was given by Townshend et al. [79], who analysed, by backscatter and transmission Raman spectroscopy, external calibration samples containing clorpheniramine maleate, and then constructed linear calibration models by means of second derivative Raman spectra, which were able to predict API concentration in tablets in agreement with HPLC results. The combination of chemometrics with Raman spectroscopy also brings advantages, as in the "non-invasive" and "non-destructive" determination of API in capsules in the presence of interfering signals in the capsules [80,81]. In the analysis of sodium diclofenac and theophylline in commercial injection solutions by FT-Raman, both multivariate (PLS and PCR) and univariate (intensity ratio) calibration methods could be applied, with a relative standard error of prediction (RSEP) only a few times smaller for the multivariate method [82].

Both NIR and Raman spectroscopy are commonly used techniques for the evaluation of polymorphs, as well as mid-infrared spectroscopy (MIR), differential scanning calorimetry (DSC), thermogravimetric analysis (TGA), X-ray diffraction (PXRD) and solid-state nuclear magnetic resonance (ss-NMR) [83]. Polymorphism is the ability of the substance to exist in multiple crystalline solid forms [65, 83,84]. Many APIs receive regulatory approval only for a single crystalline form, since polymorphs may differ on their physico-chemical properties, including melting point, density, morphology, color [83], solubility, dissolution profile, compressibility and stability. The literature offers several examples of the quantification of polymorphic forms of APIs, including the determination of imidafenacin [85] and piroxicam [86] by NIR combined with chemometric methods, as well as the quantification of two polymorphic forms of the anticonvulsivant carbamazepine by Raman spectroscopy and multivariate calibration methods [87]. Finally, sulfamerazine polymorphs were determined by several methods (DSC and linear calibration curves, X-ray diffraction and PLS, X-ray diffraction and linear regression models, Raman microscopy in mapping mode and PLS) and both the DSC/linear calibration and X-ray/PLS methods provided satisfactory results, while the X-ray/linear calibration method was not appropriate for quantification of polymorphs [88].

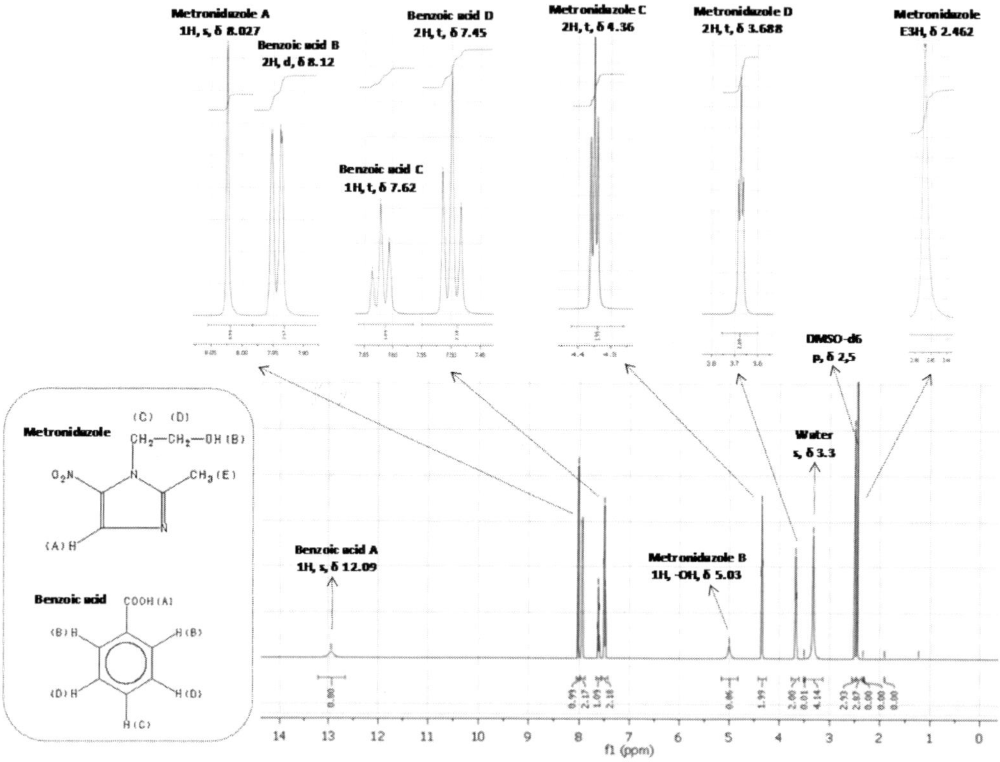

Figure 7. ^{1}H-qNMR spectrum for the API metronidazole, in presence of the internal standard benzoic acid. Experimental conditions: Solution containing 5 mg g^{-1} of metronidazole and 5 mg g^{-1} of benzoic acid (NIST, batch 350b) in DMSO-d_6;BrukerBioSpin 500 MHz, temperature: 298.1 K, spectral width: 20 ppm, number of scans: 32, relaxation delay (D1): 30 s, acquisition time: 4.096 s, number of FID points: 64 K.

Nuclear magnetic resonance (NMR), which is a technique commonly used for structure elucidation, has increasingly been used for quantification purposes by means of the so-called quantitative nuclear magnetic resonance (^{1}H-qNMR). The technique has the advantage of being a potential primary measurement method, which means a method that does not depend on a standard of the analyte being determined and is recommended for purity determination of certified reference materials according to the ISO Guides 34:2009 [89] and 35:2006 [90].

The experiments may be carried out by the addition of a fixed amount of an IS, usually a certified reference material, to a solution in a deuterated solvent containing a fixed amount of the analyte of interest. One example is given in Figure 3, which shows the ^{1}H-qNMR spectra of the API in the presence of the IS benzoic acid. It is important to remember that the IS must resonate at a different chemical shift compared to the resonances of the target analyte [91]. The ^{1}H-qNMR calculation of the analyte purity is usually estimated by Equation 18 [92], in which the analyte mass fraction is calculated in relation to the reference standard added to the solution.

$$P_a = (I_a \, . \, M_a \, . \, P_s \, . \, X_s \, . \, N_s) \, / \, (I_s \, . \, X_a \, . \, M_s \, . \, N_a) \qquad (18)$$

where I, N, M, X and P are integral areas, number of protons, molar masses, used masses and purities of analyte (a) and standard (s), respectively.

Another ^{1}H-qNMR arrangement is based on the use of an external standard, which consists of a standard reference solution in deuterated solvent added to a co-axial stem insert, located inside the NMR tubes containing analyte solution. The quantification by ^{1}H-qNMR can also be carried out by external calibration curves (with or without IS) and by standard-addition, using a specific standard compound like the pure form of the analyte instead of a non-specific internal or external standards. Other calibration methods applied to NMR are given by Bharti and Roy [93].

8.2.2. Calibration in Inorganic Analysis

Spectrometric techniques such as atomic absorption spectrometry (AAS), inductively coupled plasma optical emission spectrometry (ICP OES) and inductively coupled plasma mass spectrometry (ICP-MS) are widely used for the analysis of metals and metalloids in different types of matrices (e.g. food, water, beverages, fuels, minerals, drugs, pharmaceuticals products, clinical and biological materials) [94-103]. AAS is a robust and popular technique, used in the majority of analytical laboratories, where elemental determinations are made. The principle of this technique is based on the amount of electromagnetic energy of a particular wavelength, which is absorbed by electrons in the fundamental state of an atom. AAS requires that the analyte atoms are in the gas phase and the most common method to ensure this is the use of a flame (Flame Atomic Absorption Spectrometry – FAAS). FAAS is a rapid and precise method of analysis of over 60 elements (metals or metalloids); however, it is limited to a concentration of the orders of mg L^{-1}. To analyse samples where elements are present in trace concentration (µg L^{-1}), it is necessary to change the method of atomisation. The use of an electrothermal graphite furnace (AAS with graphite furnace – GFAAS) instead of a flame reduces the detection limit by approximately three orders of magnitude. The GFAAS and FAAS measurement principle is the same and, in this case, better sensitivity is achieved due to higher atom density and longer residence time in the graphite furnace [94, 104].

Another popular and sensitive technique is hydride-generation atomic absorption spectrometry (HG-AAS). This technique provides a degree of separation of the analyte from the matrix, minimising interferences, and consists of three steps: volatile species generation, transport of the volatile species and finally their atomisation. The detection limits of the elements determinable by hydride generation (arsenic (As), bismuth (Bi), germanium (Ge), lead (Pb), antimony (Sb), selenium (Se), tin (Sn) and tellurium (Te)) are below the µg L^{-1} range; cold vapour mercury (CV) also has extremely low detection limits due to a higher sampling efficiency. The most advanced systems combine automation of the sample introduction and hydride separation using flow injection techniques with decomposition of the hydride in an electrically heated, temperature-controlled quartz cell.

Differently from AAS, ICP OES and ICP-MS techniques use an inductively coupled plasma (ICP) source to dissociate the sample into its constituent atoms or ions. While FAAS and GFAAS have upper temperature ranges in the area of 3300 K, the plasma temperature is between 6000 and 10000 K, depending on the region in the plasma. This difference is responsible for improving excitation and ionisation efficiency and also reduces or eliminates many of the chemical interferences found in AAS [105-107].

The definition of plasma is an electrically neutral gas made up of positive ions and free electrons. Plasmas have sufficiently high energy to atomise, ionise and excite virtually all elements in the periodic table [108]. ICPs are formed by coupling energy produced by RF generator to the plasma support gas with an electromagnetic field; the field is produced by applying an RF power (typically 700–1500 W) [108]. The plasma is formed in a stream of argon gas (~15 L/min) flowing through an assembly of three concentric quartz tubes known as the plasma torch. The torch is encircled at the top by an induction coil connected to a free-running or crystal-controlled radiofrequency (RF) generator. Most the ICP's instruments operate at either 27 or 40 MHz. The Ar ICP is capable of exciting-ionising a wide range of elements, particularly metals, and it therefore allows simultaneous multi-element determination [105].

In ICP OES, the electromagnetic energy emitted by the excited atoms and ions in the plasma is measured to obtain information about the interest element concentration. The excited species in the plasma emit electromagnetic energy at several different wavelengths. Because of this, the polychromatic radiation must be separated into individual wavelengths and the emission from each excited species can be identified and their intensity can be determined without interference from emission at other wavelengths [107].

ICP-MS uses the same ICP source and sample introduction system as is used in the ICP OES. However, in this case, the ions are detected, rather than the light that they emit. The ions are extracted from the plasma and passed into the mass spectrometer, where they are separated according to their atomic mass/charge ratio (m/z) by a quadrupole or magnetic sector analyser [105–107]. ICP-MS is generally recognised as the most powerful technique for the trace detection of metals (ng L^{-1}) and exhibits a superior performance: very low detection and quantification limits, good precision, sensitivity, simultaneous multi-elemental detection and broad dynamic concentrations ranges (4–11 orders of magnitude) for many elements [94, 105].

8.2.2.1. Methods of Quantification

The majority of analytical methods do not directly provide the concentration of elements in an unknown sample; the consequence of this limitation is the need for a calibration procedure. There are many different methods of quantification to carry out metals and metalloids analysis by spectrometric techniques (AAS, ICP-OES and ICP-MS), depending on the technique utilised, the required data quality, accuracy and precision. Therefore, the analyst will decide on the best strategy that applies to his samples: qualitative, semi-quantitative and quantitative analysis [105, 106, 108, 109].

The qualitative analysis is the investigation of the presence (or absence) of a metal/metalloid or group of these metals/metalloids in a sample. This method is the simplest way to obtain initial information about a sample, and is generally employed to totally characterise an unknown sample.

The qualitative analysis can be applied to techniques such as AAS, ICP OES and ICP-MS. When the AAS is used, special systems of AAS are needed, simultaneous or sequential multi-elementals systems. The ICP OES is able to simultaneously analyse elements in a wide wavelength range from ~130 nm to 850 nm with the best spectral resolution, while the simplest mass spectrum for qualitative analysis in ICP-MS is normally obtained by using a spectral scanning procedure from m/z value 1 to 240; the spectrum consists of a file with a single data point at each mass.

ICP-MS also offers excellent semi-quantitative capabilities [108, 105, 106, 108, 109]. This method of analysis is carried out by the response of a detector from different masses along the spectrum, resulting in a curve. The profile curve can be obtained using 6–8 different mass elements. The response of each element is adjusted according to the isotopic abundance, concentration and ionisation efficiency. The response curve is saved and redefined daily before starting the analysis, since its profile depends on the instrument optimization conditions. The multi-elemental semi-quantitative method provides an accuracy of about ± 50% or better for the amount of each element present.

Quantitative analysis is used when high accuracy and precision on the results are required. To obtain high accuracy, interferences must be absent from the analysis procedure or strategies provided for their correction. The main methods applied in quantitative analysis are external calibration, standard addition, internal standardization and isotope dilution [105, 108]

(1) External Calibration

The external calibration approach is usually used when the difference between the matrix sample and standard solution is negligible. It is a general method for determining the concentration of an analyte in a sample using successive and increasing standard solutions plotted versus signal intensities of these standards, generating thereby the calibration curve.

This curve establishes a relationship between the metal concentrations on the standard solution with the signal intensity, which generally is linear.

The ordinary least squares (OLS) regression is the most common approach used on commercially available instruments of atomic spectrometry and the evaluation of the regression adjustment is restricted to the determination coefficient, r^2.

In AAS, ICP OES and ICP-MS, it is relatively easy to obtain r^2 greater than 0.999, but it is necessary to remember that this criterion is not enough to ensure the adjustment of the regression model [110-112]. In this way, other parameters should be also evaluated to assure that the equation obtained by the OLS method is able to be applied for sample analysis with reliable results. The parameters that the analyst should evaluate in a calibration curve are: the residuals behaviour, the linearity (which is considered one of the most important characteristics for result accuracy [113]), the lack of fit and pure error of the model. Besides these parameters, Mark and Workman [114] proposed that the statistical significance of the regression coefficients might be evaluated as a good estimative of the best curve fitting.

(2) Standard Addition

This method of calibration provides an effective way to minimise matrix effects by spiking samples with known concentrations of analytes. The main advantage of the standard addition method is that the calibration curve is built with the matrix on the unknown sample. This procedure consists of using three or more known crescent quantity spikes of the analyte with equal aliquots of the sample. The signal intensities of the spike addition sample and an unspiked aliquot of the sample are plotted versus the quantity of the analyte spiked on the samples. The fit of these points results in a linear function and the point at which the extrapolated curve intercepts the abscissa (negative side x-axis) indicates the quantity of the analyte in the unknown sample [108, 109]. This process is more laborious and more time-consuming when compared to the external calibration method and, due to this, it is generally limited to small numbers of analytes and samples, and is not suitable for multi-element analysis.

(3) Internal Standardization

The internal standardization method is used to obtain high-quality data on the calibration curves. This method is employed to correct changes in analyte sensitivity caused by variations in the concentration and the type of matrix compounds found in the sample [109], and to correct instrumental fluctuations during analysis. By using appropriated ISs, the stability can be improved by an order of magnitude, which is reflected in the overall accuracy of the quantification [108].

In ICP-MS, an IS is a non-analyte isotope that is added to the blank solution, standards and sample, in the same amount, before analysis [109]. The efficiency of this IS requires that its behaviour is as close as possible to the elements analysed. Its choice is limited by the presence of this element in the natural sample and must not interact with analytes from the sample. If ICP-MS is the technique employed on the analysis, the IS must be free of isobaric interferences and polyatomics.

After acquiring the signal intensities, mass spectrum and the ratio of the analyte signal intensity to the IS signal intensity are plotted versus the known concentration of the analyte in the calibration standard; the least squares regression equation is obtained for these data.

(4) Isotope Dilution (ID)

Isotope dilution is an absolute method of quantification with high accuracy and precision, and provides definable uncertainty values. This analytical method is based on the measurement of the isotope ratio in samples, where isotopic composition (natural abundance) is altered by the intentional addition of a known amount of an enriched isotope of the same element (spike). The analysed element needs to have at least two stable or long-lived radioactive isotopes able to be measured in a mass spectrometer free of spectral interferences [115-117].

In practice, basically, this calibration procedure involves some main steps [118], as described below.

Initially, one must estimate the concentration of the analyte in the sample by ICP-MS or one of the alternate methods:

(i) the sample is gravimetrically spiked with an isotopic analogue of the analyte, forming a sample solution blend;
(ii) a calibration standard is gravimetrically prepared with concentrations of natural analyte and isotopic analogue, which are the same as those estimated in the spiked sample solution blend; and, using appropriate ions of the natural analyte and the spike, the isotopic amount ratios for the spiked sample and the spike calibration are determined. The mean value of the five or more measurements will minimise the effects of an instrument drift. An improved estimate of the natural analyte concentration in the sample can then be calculated from the data.

The values of the isotope amount fraction ration can be calculated from certified values. The experimental value of the ratio for the calibration standard is usually determined for each sample by analysing it before and after the unknown sample is measured.

The polyatomic and oxides interferences must be minimised by the tuning instrumental parameters. In the same way, the presence of isotopic interferences must be checked by comparing spiked and unspiked standards or samples [118].

After having achieved equilibrium of the spike with the sample solution blend, the modified analyte isotope ratio is measured and calculated by Equation 19 [115]:

$$c_s = c_{sp} \frac{m_{sp}}{m_s} \frac{M_s}{M_{sp}} \frac{R_b A_{sp}^b - A_{sp}^a}{A_s^a - R_b A_s^b} \quad (19)$$

where:

c_s and c_{sp}: sample concentration of the element in the sample solution blend and spike solution;

m_s and m_{sp}: mass of the sample solution blend and spike solution;

M_s and M_{sp}: molar mass of the sample solution blend and spike solution;

$A^a{}_s$ and $A^a{}_{sp}$: amount fraction of isotope “a” sample solution blend and spike solution;

$A^b{}_s$ and $b^a{}_{sp}$: amount fraction of isotope “b” sample solution blend and spike solution;

Rb: measured isotope ratio of the sample solution blend corrected for mass discrimination (CF).

Where:

CF = t/e

t = true ratio of the isotope amount fractions calculated from certificate values

e = experimental ratio of the isotope amount fractions obtained using a mass spectrometer

The main advantages of isotope dilution are [115, 118]:

(i) once equilibrium of the spike and sample solution blend has been achieved, the total recovery is not required, because the final value is based on measuring the isotope ratio between the analyte and the spike; eventual losses of the analyte do not affect the results, since both isotopes will be affected in the same way;

(ii) the accuracy of the method is determinate determined by the precision of this ratio measurement; and the methodology is less time-consuming and can provide greater accuracy than standard additions.

And the main disadvantages are: the high cost and availability of isotopic materials; it is a destructive method, the cost of mass instrumentation; training the analyst is required; and the requirements of isotopic equilibrium.

8.2.2.2. Application of Calibration in Atomic Spectrometry Techniques

OLS regression is the most common calibration approach applied in atomic spectrometry. Multivariate calibration is an emerging approach and its application is more frequent on complex atomic interference or in analysis with complex interferences.

In this section are presented the calibration approaches that have been applied in the analysis of samples such as environmental, food, fuels, biological, etc., by atomic spectrometric techniques (AAS, ICP OES and ICP-MS).

(1) Atomic Absorption Spectrometry

Classical calibration procedures (external calibration, addition standard, internal standardization for quantitative analysis) are the most common strategies used to carry out measurements by AAS techniques.

The choice of the calibration procedure will depend on the presence of interferences and the concentration of the analyte.

Penninckx et al. describe a complete procedure for validation of the calibration methods in AAS techniques. The authors gave an overview of the parameters that should be considered in evaluating the performance of the calibration approach. The example of determination of zinc (Zn) and iron (Fe) by FAAS and Pb, cadmium (Cd) and manganese (Mn) by ETAAS are used for illustration. In this work, it was demonstrated that, depending on the range of the calibration curve, the variance of the data set may have a homoscedastic or heteroscedastic behaviour. In the case of non-constant variances or non-linearity, it must be checked whether a straight-line model can be valid over a smaller concentration range. This procedure allows the use of an unweighted instead of a weighted model. Another option is to investigate the suitability of a second degree model or use a weighted model [119].

Husáková et al. determined arsenic (As) in beer by ETAAS using a univariate optimization to eliminate the interference of phosphate on analytical signal and choose the most adequate chemical modifier and pyrolysis and atomisation temperatures. The authors used the external calibration approach and validated the method by analysing samples of certified reference materials (CRM) [120].

Certified reference materials are an important tool in the validation of method and are generally used to evaluate the accuracy of methods [121,122]. These materials can also be used to identify systematic error due to the strategy of calibration or sample preparation.

Yaman carried out a complete study of optimization and validation for determination of Pb and Cd in samples of tissue by FAAS. In this work, the univariate method was used to optimise the method and external calibration was used for quantifying the amount of Cd and Pb. The correlation coefficient was used to evaluate the linearity and figures of merit such as selectivity, precision and accuracy were reported [123].

Internal standardization was used by Caldas et al. to take into account the matrix effect on the measurements of As, Pb and copper (Cu) in sugar-cane spirits by GFAAS. Univariate study was conducted to optimise the method and the linearity was evaluated through the r value [124].

Lyra et al. studied a procedure for determining sodium (Na), potassium (K), calcium (Ca) and magnesium (Mg) in biodiesel by FAAS using microemulsion as sample preparation. The authors compared the results obtained from the aqueous and organometallic standard in order to verify if the calibration method has the same answers when the metals are in the "free" forms or bonded to organic molecules. External calibration and standard addition were compared. The results showed that the atomisation of the analyte in organometallic standard solutions is not affected when the matrix is present. The r^2 was used to evaluate the linearity, but the authors did not report the calibration range [125].

Pereira et al. determined Cd and Pb at low levels in mineral and drinking water. The preconcentration was carried out by an on-line system, using ammonium pyrrolidinedithio-carbamate (chelating agent) and fullerene as adsorbent. The measurements were carried out by thermospray flame furnace AAS (TS-FF-AAS). The authors used OLS regression to evaluate the linearity of the dataset and reported the *r* value greater than 0.999. The accuracy test was carried out using a CRM of water [126].

Preconcentration and separation procedures are generally used to reduce the effect of interferences and improve the analytical signal.

There are two main classes of interference that can affect the measurements in AAS. The first is related to the spectral interference caused by other components present in the sample and the interferences caused by chemical processes occurring in flame and electrothermal atomisers. The effect of chemical interferences is usually more important than spectral interferences in ETAAS, where its effect is more pronounced [127].

The analysis of environmental samples, different from water, is more complex due to the matrix composition and chemical interference is more common. In ETASS, the use of chemical modifiers is required to correct for chemical interferences that originated due to the high amount of matrix concomitants introduced into the electrothermal atomiser. The optimization of the procedures used in ETAAS is not simple and is time-consuming [127]. Sometimes, spectral and chemical interferences cannot be totally eliminated using chemical modifiers and some researchers are making use of multivariate calibration to circumvent these problems.

Felipe-Sotelo et al. used PLS multivariate regression in order to take into account the spectral and chemical interference of Fe on chromium (Cr) determination in water by ETAAS [127]. The direct determination of Cr by ETAAS is considered difficult because the deuterium lamp has too low emission intensity at the resonance line of Cr (359.9 nm) to ensure satisfactory background correction [128]. The PLS model was able to predict the concentration of Cr, at low level, when another concomitant metal causing spectral or chemical interference was present. The performance of the calibration model was tested by analysing CRM. With the multivariate approach, the authors have eliminated the use of chemical modifiers and avoided preconcentration and separation steps. They demonstrated that linear PLS was able to predict three commons effects that currently occur in ETAAS: peak shift, peak enhancement and random noise [127].

Grotti et al. determined Mn, Cd and Fe in seawater by ETAAS. Seawater presents high levels of Ca, Mg, Na and K as nitrates and these salts affect the atomisation of Mn, Cd and Fe. Four different calibration approaches were compared in the analysis of these transition metals: external calibration, multivariate calibration and external calibration using matrix-matched standard and addition standard. CRM were used to evaluate the performance of calibration method and Student's t-test was used to compare the certified value and results obtained. The multivariate calibration presented the better results among the approaches applied. The results obtained from external calibration using the matrix-matched standard and addition standard were unable to correct the effect of alkaline and alkaline earth metals on Cd, Fe and Mn and systematically yielded overestimated results [129].

Vieira et al. quantified the amount of As in samples of sediments, coal and fly ash slurry by hydride generation atomic absorption spectrometry. The results obtained from external calibration and addition standard were compared and the accuracy was evaluated using CRM. The results obtained with external calibration were significantly lower than the certified values and the authors concluded that the arsine production was more effective in the aqueous standard. The addition standard was employed to reduce the matrix effect and the results were improved and no significant difference was detected at a confidence level of 95% [130].

Felipe-Sotelo et al. used multivariate calibration to determine Sb in samples of soils, sediments and coal fly ash by ETAAS. The authors used this method to model the matrix effect and the background corrections on the measurements. Instead of using the univariate optimization of the method, the authors used experimental designs at different levels of Sb in order to develop the calibration matrix and identify which concomitants had an influence on the analytical signal. The method was validated with five CRMs and the main limitations of the models were related to the mass/volume ratio and the low amount of Sb of the solid matrices. The PLS approach gave good results and was able to identify the problems of non-linearity caused by peak displacement, peak enhancement/depilation and peak tailing. The methodology developed by the authors does not require the use of chemical modifiers and reduces the use of reagents and time [131].

In a similar work, Petrucelli et al. applied the multivariate calibration models based on PLS to determine Cd and Pb in sewage sludge samples by TS-FF-AAS. The authors compared three different calibration strategies. Once more, the PLS models presented better results when compared to external calibration. Furthermore, by using the multivariate calibration, it was possible to improve the method sensitivity (LOD and LOQ were one third lower than the values obtained in linear calibration) [132].

Baxter et al. used the PLS method to solve the spectral interference of aluminum (Al) on the measurement of As on marine sediments by ETAAS. In this case, it was not possible to use an alternative wavelength due to the low concentration of As. The authors demonstrated that the addition standard method was unable to correct the spectral interference because of the interaction between Al and As [133].

(2) Inductively Coupled Plasma Optical Emission Spectrometry

Escudero et al. determined Zn in natural water by ICP OES using a solid-phase cartridge for on-line sample concentration. The authors used the univariate method to evaluate the effect of pH, sample flow rate and concentration of eluent and buffer solutions on the sorption conditions. External calibration was used and the r^2 value was considered for evaluating the linearity. Validation was carried out using a CRM of water [134].

Mutic et al. proposed a method for determination of As in wine by ICP OES using internal standardization as the calibration strategy. As the IS, the authors compared four elements, cobalt (Co), bismuth (Bi), vanadium (V) and yttrium (Y), and the accuracy of the method was checked using a CRM of groundwater. When the results of the CRM obtained from external calibration were compared to internal standardization, no significant difference was detected. However, the spike recovery studies in wine demonstrated that the matrix has a significant effect on the external calibration performance with recoveries smaller than 65% [135].

Souza et al. compared two kinds of micronebulizers for elemental determination in lubrificant oil, residual oil and biodiesel by ICP OES. In this work, the method was optimised by a factorial design with replications at the central point. The external calibration approach was used and the accuracy of the method was evaluated using CRM residual fuel oil and lubrificant oil [136].

Considering that metals present in vegetable seeds are a source of contamination of biodiesel, Chaves et al. determined the amount of these metals and phosphorus (P) in different kinds of vegetable seeds used in the production of biodiesel by ICP OES. Samples digestion was carried out in a microwave oven. External calibration was used to evaluate the dataset and the authors found r^2 values greater than 0.99. A CRM of pine needles was used to evaluate the accuracy of the method [137].

In a similar work, Naozuka et al. developed a method for analysing ten metals, Al, barium (Ba), Ca, Cu, Fe, K, Mg, strontium (Sr) and Zn, two metalloids (boron [B] and siliceous [Si]) and three non-metals (chlorine [Cl], P and sulphur [S]) in nuts and seeds by ICP OES. The samples were decomposed in a closed vessel microwave digestion system. The r^2 value was used as the criterion to evaluate the linearity of the calibration curves. For some analytes, the calibration curve ranged from 0.5 to 500 mg L^{-1}. The method accuracy was verified by analysing the CRM of peach leaves. The authors concluded that the results were in agreement with the certified value at a confidence level of 95% [138].

In a recent paper, Mermet describes a tutorial review about the weighting procedure for calibration in atomic spectrometry [139]. It was demonstrated that when using a large range of concentration, a non-weighted regression is mainly useful in the upper mid part of the range while the weighted regression model is more adequate in the lower part. Additionally, it was suggested that the uncertainty of the regression is more adequate instead of the coefficient correlation for the selection of the most appropriate regression model [139].

Despite the existence of recommendations for applying the WLS regression model when using a larger range in the calibration curve, Chaves et al. did not provide additional information related to the variance of the dataset or if the weighing procedure was performed [137]. In most of the cases that involve atomic spectrometry methods, the OLS regression model is preferred instead of the WLS model because the last is a more laborious procedure.

A practical example of the effect of weighting procedure on the analytical results is given by Nascimento et al. in the determination of metals in human milk by ICP OES. The developed method was validated and the accuracy was verified using a CRM of infant formula. For some metals such as V and Zn, the concentration of the calibration curve ranged from 1 to 100 mg L^{-1} and this is a typical case where it is necessary to verify if all the conditions for using the OLS regression are fulfilled. During the validation of the method, the authors performed a study aiming to compare the performance of the OLS and WLS regression models. The paired Student's t-test was used to compare the angular coefficient obtained from the OLS and WLS regression models. As examples, when comparing the analytical curve derived from OLS and WLS regression for K, it was observed that a significant difference between both models exists at all ranges, while for Zn, this difference is significant only at high concentrations [140].

Griffiths et al. compared the traditional and multivariate calibration techniques for determination of platinum (Pt), palladium (Pd) and rhodium (Rh) in autocatalyst samples. The PLS method presented superior performance when compared to the univariate methods (single-point calibration, interelement correction and matrix-matched calibration) [141].

Pimentel et al. applied PLS and PCR to determine simultaneously Mn, molybdenum (Mo), Cr, nickel (Ni) and Fe in steel samples, using low-resolution ICP OES with diode array detection. The model takes into account spectral interferences and the errors of measurements were slightly smaller than those obtained with univariate calibrations [142].

The investigation of interferences in ICP OES is generally carried out using high power plasma and the use of lower power plasma is favourable in situations where it is necessary to increase the signal-to-background ratio and consequently reduce the detections limits. Plasma robustness is a critical condition to minimise matrix-induced changes in the plasma source. In order to quantify the interferences caused by the matrix effect on the signal-to-background ratios in ICP OES measurements, Grotti et al. made a study using a multivariate calibration approach aiming to take into account the effects of Na, K, Ca, Al and Fe at different concentration levels. The authors used these data to optimise the method and achieved the conditions of plasma robustness at low power plasma [143].

In a practical example, Kola et al. utilised a multivariate regression to minimise the interference effects caused by Ca in the S measurements by ICP OES. The Ca causes spectral overlap with S and one way to correct this interference is to calibrate the instrument first with pure S standards, then measure a series of Ca standards as samples and then fit the regression line so that the apparent S concentration is a function of the Ca concentration. This correction procedure is time-consuming and requires additional work. Using the multivariate regression, Kola et al. developed an efficient method for correcting the spectral interference caused by Ca on S measurements. The main advantage of this method is the reduction in consumption of standards and time [144].

(3) Inductively Coupled Plasma Mass Spectrometry

Semi-quantitative analysis by ICP-MS is commonly used for fast screening purposes. The reliability of this method was evaluated by Chen *et al.* The authors compared the semi-quantitative method with the external calibration method using certified reference values of CRM of natural or fortified fresh water. Better accuracy and reproducibility were obtained when used multi point calibration but they concluded that the semi-quantitative analysis mode could be a reliable and robust method in situations where rapid screening is required or limited sample information is available [145].

Gerdes and Carter developed a calibration strategy for semi-quantitative analysis of gas phase samples using an ICP-MS equipped with an Octopole Reaction System (ORS). The calibration approach used a single tracer isotope in liquid phase. The method presented accuracy of 50% for the most elements in the low mass range and 20% in the middle and high mass range using the isotope cesium 133 as a reference isotope. The method was able to quantify metal concentration lower than 1 ppb in gas phase [146].

Rocha *et al.* determined trace elements in edible seaweeds by ICP-MS. They quantified As, Cd, Co, Cr, Mo, Ni, Pb, Sb and V and for validation of the method CRM of sea lettuce, tea, and peach leaves was used. For calibration, four multi-elemental solutions were used and the linearity was assessed by the correlation coefficient and as criterion r values greater than 0.9995 were considered an indication of the model adequacy. The results obtained from CRM presented mean deviation of 5% and the authors considered the method adequate for the analysis [147].

Ataro *et al.* quantified trace elements in raw cow's milk by ICP-MS. The method was developed using external calibration and data from recoveries using CRM of whole milk powder was utilized for evaluating the method performance [148].

Hipper *et al.* compared two methods for the determination of methyl mercury in whole blood samples by ICP-MS coupled to a gas chromatography and gas chromatography-mass spectrometry. The authors used the external calibration to quantify the analyte and the r value was used as criterion to evaluate the linearity of the model [149].

Monitoring of heavy metals in drug substances in an important activity n pharmaceutical industry and the ICP-MS technique is the better choice due its low limit of detection.

Suzuki *et al.* determined Na, Pd, Ba, I and Br in methamphetamine by ICP-MS. They presented the calibration graphs and concluded that linearity was good without presenting-any statistical data to support this conclusion [150].

Lewen *et al.* have used the ICP-MS to determining Pd in fosinopril sodium. Indium was used as an internal standard and the linearity of the method was 2.5 to 50 ng mL^{-1}. The coefficient of correlation was used to evaluate the linearity and in addition was carried out recovery studies [151].

Dolan *et al.* analyzed samples of dietary supplements for determination of As, Cd, mercury (Hg) and Pb using ICP-MS. They used a high-resolution ICP-MS in order to avoid isobaric interferences. Internal standard was used aiming to improving the accuracy of the method. The method was validated using CRM of orchard leaves, typical diet, whole egg powder, bovine liver, and trace elements in spinach, bone meal. The authors used the recovery test to evaluate the procedure of samples preparation and quantification [152].

Pick *et al.* described the use of the dynamic reaction cell (DRC) for suppressing or reducing polyatomic interferences in the analysis of biological material by ICP-MS. The authors determined As, Cr, Fe, Ni and selenium (Se) in plants and animal feedstuffs. It was used the standard addition calibration approach and internal standard correction for minimization of the matrix-induced interferences (non-spectral interferences). The accuracy of the method was evaluated by using three CRM (a compound feed, and animal feedstuff for piglets and Lucerne). The use of DRC led to lower detection limits and improved the accuracy [153].

In the similar way, Guo*et al.* have studied the elimination of oxide interferences and determined ultra-trace silver (Ag) in soils by ICP-MS with ion-molecule reactions. Accurate determination of low levels of Ag in the presence of zinconium (Zr), niobium (Nb), Y and Mo is difficult due the formation of oxides and hydroxides, whose isotopes overlap those of Ag.

In order to eliminate the interference caused by oxides and hydroxides on Ag determination, they have adopted a strategy based on ion-molecule reaction with oxygen (O_2) as the reaction gas into DRC. After univariate optimization, an analytical curve with concentrations from 0.05 to 50 ng mL^{-1} was used and the correlation coefficient (greater than 0.999) was used as parameter to evaluate the linearity. The method was validated using CRM of soil [154].

Isotope dilution is a quantification method that has been increasingly applied in the bioinorganic speciation field, especially in the case of organotin and organomercury. This method is more accurate and precise than the standard additions method.

Mcsheehy *et al.* determined methionine and selenomethionine in selenium-enriched yeast by isotope dilution with liquid chromatography- mass spectrometry and ICP-MS. The results were compared with GC-MS reference method developed in-house by the authors and it was concluded that the results were similar [155].

Ng *et al.* determined trace elements in drinking water by exact matching isotope dilution mass spectrometry (ID-MS). They determined the concentration of Ni, Cd and Pb with uncertainty of less than 2%. It was used CRM of trace elements in water as an internal quality control [156].

Spectral and non-spectral interferences are a common problem in analysis with ICP-MS. In order to eliminate or reduce this problem the addition standard method, DRC or high resolution equipments are largely used. The chemometric techniques for calibration modeling are generally used in low resolution ICP-MS in order to take into account the interferences.

Donachie *et al.* applied the multivariate calibration using electrochemical/ICP-MS for determination of trace elements in water and humic acid matrices. They compared the techniques of MLR, PCR and PLS. The MLR was not creating regression coefficients which had the ability to model the full span of the data space and problems with the inverse matrix and this affected the collinearity. PCR described the data well but there were still errors while the PLS model gave the most accurate overall predictions and all the concentration at the lower levels [157].

Rupprecht and Probst applied different multivariate regression method to corrected spectral and non-spectral interferences in ICP-MS measurements. They compared the performance of PLS and PCR methods with OLS regression. The combination of internal standard and multivariate calibration methods offered the possibility to correct for spectral and non-spectral interference in ICP-MS analysis [158].

Moberg *et al.* determined Cd in fly ash and metal alloy in five CRM. Analysis of Cd in these types of matrices is problematic due the spectral overlap for all analytically important Cd isotopes. The authors used the chemometric methods to overcome this problem and to take into account the interferences. The univariate calibration overestimated the results due the high interferences caused by Mo while the multivariate approach gave results similar to the certified values [159].

8.3. Calibration Methods Applied to Polymeric Materials Characterization

Polymeric materials are an important class of organic compounds and various experimental techniques involving calibration methods are employed to study their properties. The most used techniques where calibration is necessary for the characterisation of polymeric materials are Raman spectroscopy [160], FTIR (Fourier transformer infrared spectroscopy), NIR (Near-Infrared) spectroscopy [161], X-ray Fluorescence [162,163], Time-of-Flight Secondary Ion Mass Spectrometry (TOF-SIMS) [164] and Nuclear Magnetic Resonance (NMR) [165].

Works about calibration methods applied to polymers are used to clarify some specific properties of this class of materials, including: degree of crystallinity [166], additives concentrations [167], composition of processed polymer or blend [168], moisture content [169], status of reaction in a reactive extrusion [170], presence of recycled polymers [171] and quantification of heavy metals [172].

The presence of chemical groups in polymeric materials can be deduced through vibrational spectra [173]. Band intensities have quantitative information about these chemical groups, and a relationship between band intensities and concentration of the mixture components can be achieved using a calibration model. This case is possible with Raman spectra; however, in NIR spectra, the overlapping of bands and intensities can cause difficulty in the construction of a univariated model. In this situation, it is necessary to use multivariate methods to extract all the information from these spectra. The most used chemometric methods to analyse spectral data of polymeric material are principal component analysis (PCA) and its quantitative models, such as principal component regression (PCR) and the partial least square (PLS) method.

8.3.1. Calibration Methods Applied to In-line Monitoring of Extrusion Process

Chemometrics analyses can also be used to determine the composition of a polymer blend and additives during the extrusion process. Due to increased concern about the environment and the health of consumers, the real-time monitoring of chemical reactions that occurs during the extrusion step has become crucial for the plastics industry. Using a suitable calibration model, it is possible to predict real-world samples in real time, during a processing step.

Fischer et al. [174] monitored the polymerisation processes and the composition of polymer blends and additives during extrusion processes. This paper was based on four examples of in-line spectroscopic monitoring of the extrusion process applying calibration/validation methods: in the first example, a model was constructed to detect the modification of OH end groups of hyperbranched poly(urea-urethane)s with ATR-IR spectroscopy, the second example was to realise the observation of the synthesis of a halogen-free amino-functionalised poly(methyl methacrylate) (PMMA) using Raman spectroscopy, the third example was to realise the real-time determination of an additive content, a flame-retardant melamincyanurate (MC) in polyamide-12 (PA-12), during the extrusion process using NIR spectral data and the last one was to determine the composition of a PE/PS (polyethylene/polystyrene) blend during extrusion using two techniques, NIR and ultrasonic spectroscopy. Here, the third and fourth examples (quantification of the flame-retardant MC in PA-12 by NIR spectroscopy and quantification of the composition of a PE/PS blend using NIR and ultrasonic spectroscopy) will be discussed in more detail.

Polyamide is an engineering plastic with innumerable industrial applications [175]. Its flammable character makes necessary the use of flame retardant in a percentage between 5 and 16 wt%, depending on its application. For the calibration, 21 different PA/MC formulations (126 spectra, 6 for each formulation) were prepared and 3 formulations were used for independent validation. The PLS method was used for three different data pretreatments: without pretreatment, multiplicative scattering correction (MSC) [176][1] and standard normal variate (SNV) [176][2]. The PLS procedure used the three main components and the correlation coefficient (r^2) obtained by the three tested pretreatments was 0.99. The authors also reported an error of cross validation of 0.06% for the three calibrations.

The real-time monitoring of the composition of polymeric blends during extrusion is important because variations in concentrations of one of the components can change the mechanical properties of the blend, as impact strength. Using the PLS model, an r^2 of 0.997 (six main components) and a standard error of 0.97% were observed. The validation with two

[1] In the Multiplicative Scatter (or Signal) Correction (MSC), the light scattering or change in path length for each sample is estimated relative to that of an ideal sample. In principle, this estimation should be done on a part of the spectrum that does not contain chemical information, i.e. influenced only by the light scattering. However, the areas in the spectrum that hold no chemical information often contain the spectral background where the signal-to-noise ratio (SNR) may be poor. In practice, the whole spectrum is sometimes used. This can be done provided that chemical differences between the samples are small. Each spectrum is then corrected so that all samples appear to have the same scatter level as the ideal. As an estimate of the ideal sample, we can use, for instance, the average of the calibration set.

[2] Standard normal variate (SNV) pretreatment removes the slope variation from spectra caused by scatter and changes of particle size. This correction is applied to each spectrum individually by subtracting the mean spectrum and scaling with the spectrum standard deviation.

mixtures has a prediction error of approximately 1%. The quantitative evaluation of the ultrasonic values results in a prediction error for the correlation of ultrasonic damping and concentration of 1.9% and for the correlation of ultrasonic velocity and concentration of 2.2%.

Rohe et al. [168] used in-line NIR spectroscopy during the extrusion process of a polyethylene/polypropylene (PE/PP) blend, for the determination of the polymer melt composition. Fiber optic probes were used to transmit light through the polymer stream. Spectra were collected before the extrusion, as a reference spectrum, and after the extrusion. Such parameters of processing as melting temperature, melting pressure and torque of screw also were registered. For calibration development and validation, 20 NIR spectra of each PE/PP mixture were collected. For spectra statistical analysis, PLS analysis was used. The PLS model was applied for prediction of the polymer mixture, i.e. the ratio of PE to PP in 11 different PE/PP blends. PLS was used with a spectral variance of 95% and a compositional variance of over 95% of the polymeric blend PE/PP.

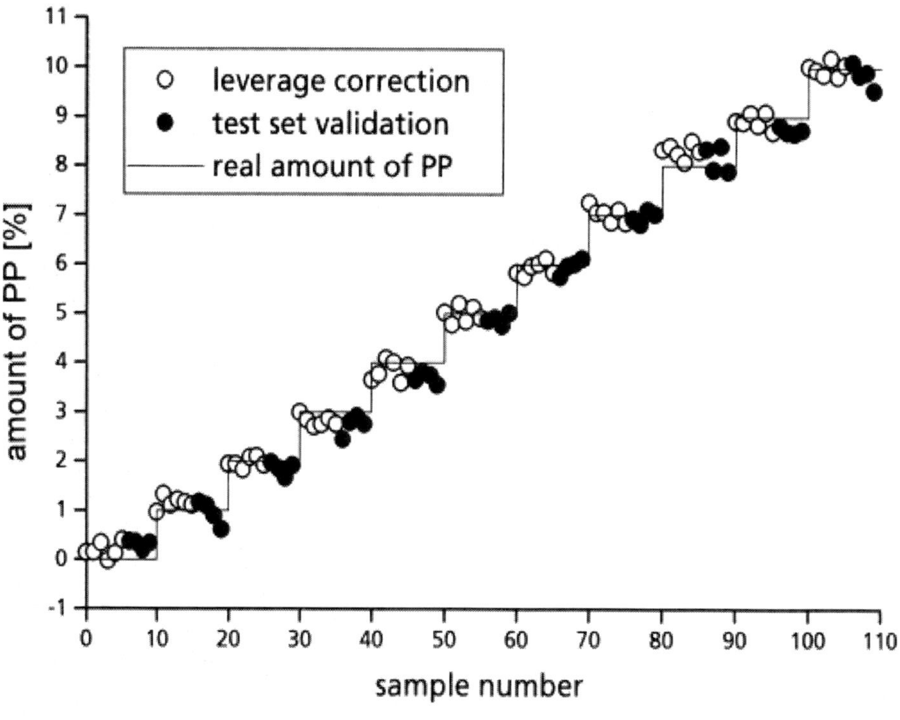

Figure 8. Comparison between the test set validation method using PLS model, validation of PLS using leverage correction and real percentages of PP in a PE/PP blend prepared by single screw extrusion.

The combination of different spectra preparation techniques was tested to generate an optimised PLS model. The test set validation was performed using six samples for calibration and four samples for validation. For this condition, a root mean squared error of prediction (RMSEP) of 0.21% was obtained, which indicates an error of ± 0.21% in prediction. Figure 8 shows a comparison between the PP content obtained by validation of the PLS model, using leverage correction and the real content of PP in these blends. As expected, using the leverage correction, the values were closer to the real values because, in this situation, the samples had already been used for the construction of a PLS model. However, both methods, test validation and validation using leverage correction, formed “clusters” around the real values of PP content.

Leverage correct validation is a simple validation method that requires only one model for validation of a final model. It gives the same scores and loadings as the validated test set, since all the samples are used to make the model. The leverage correction can be used at the initial modelling stages, but should be replaced by a more sophisticated method for final validation because, in general, the prediction of error will be lower than the real value [176].

8.3.2. Quantitative Analysis of Polymer Additives

Polymeric materials have in their formulation some chemical compounds as catalyst residues that come from the synthesis process. During the processing step, some additives are added to the polymer being melted to avoid degradation, improve the processability and maintain some interesting properties such as mechanical resistance, thermal stability and transparence [179]. Thus, FTIR spectral data of polymers is a suitable technique to evaluate the additive concentration and the appearance of degradation products. These spectral data can be easily interpreted when a calibration method, such as PLS or PCR, is used. In this context, the variable selection method is able to remove non-informative variables, producing models with good predictive capability. The genetic algorithm is an efficient method to select variables for multivariate calibration.

An example is the work of Leardi et al. [178] that combined variable selection, using a genetic algorithm, with PLS to predict the additive concentration in polyethylene films using FTIR spectra. In this work, the concentrations of three additives (A, B and C) were determined in 319 samples. For the model construction, 48 calibration samples and 28 validation samples were prepared, with a concentration range of 475 to 912 ppm, for the additive B, and 109 calibration samples and 65 validation samples, with a concentration range of 269 to 1720 ppm, for the additive C. Additive A was not analysed using genetic algorithm because there were very few samples containing this additive. Using a genetic algorithm (GA) with only a few samples, *overfitting* [179][3] can occur. The comparison between the predicted values and true values to the two tested additives (B and C) are showed in Figure 9.

The model was simplified by removing the samples that contained more than one additive. The authors compared the variables selected by the genetic algorithm with the regions of FTIR spectra selected by experts, who were familiar with the spectroscopy of these additives. For the three tested additives, the RMSEP values were similar for the GA and experts’ values, indicating that the two models have equal predictive ability. An advantage of GA variable selection, reported by these authors, is the appearance of additional bands in the spectra that sometimes are easily interpreted. These additional bands can have relevant information since they decrease the RMSEP and were selected in independent GA runs. Another advantage of the GA variable section method, used to construct reliable calibrations models, is that this method does not require any spectroscopic experience by the user.

---

[3] Overfitting occurs when a statistical model describes random errors or noise instead of the underlying relationship. Overfitting generally occurs when a model is excessively complex, such as having too many parameters relative to the number of observations. A model that has been overfit will generally have poor predictive performance, as it can exaggerate minor fluctuations in the data.

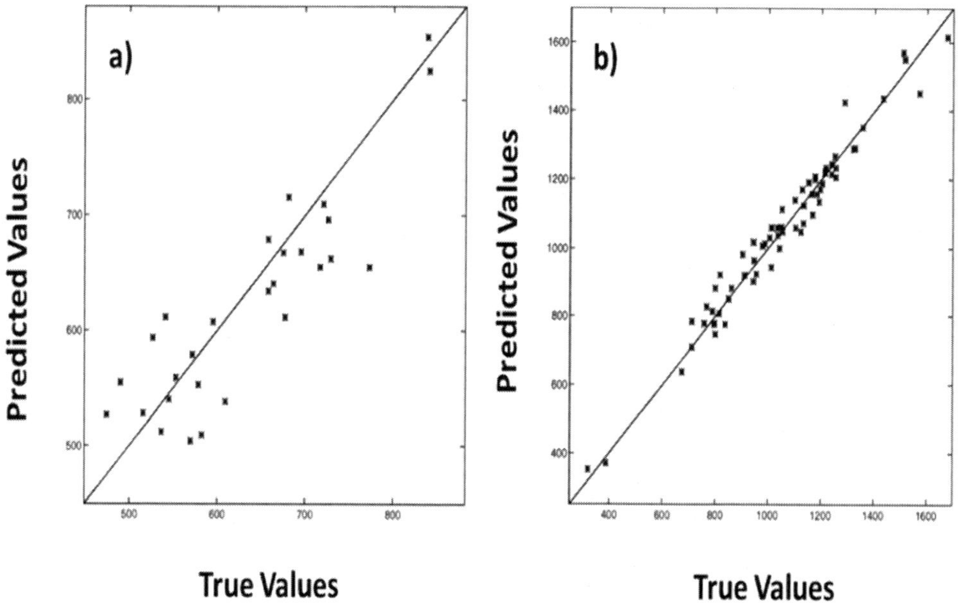

Figure 9. Predicted vs. actual concentration plots for the validation data for additives B (a) and C (b).

The detection and quantification of additives at polymer surfaces using calibration and validation methods also was treated by Bertrand et al. [179], but in this case, the experimental technique used was TOF-SIMS. Ultra-thin polymer films (< 1 μm) containing antioxidants were prepared by spin coating on silicon wafer. The polymer samples chosen were poly(ethylene terephthalate-ethylene isophthalate) random copolymer and poly(methyl metacrylate) (PMMA). They were chosen because they are amorphous, mechanically stable and can be prepared as an uniform thin film. The antioxidant used was Irgafos 168^{TM} from Ciba, or tris(2,4-di-tert-butylphenyl) phosphite (Figure 10). Irgafos 168 is a secondary antioxidant that acts as a suppressor of hydroperoxide groups.

Irgafos 168 antioxidant was chosen because this additive presents an intense and easily detectable peak in the high-mass range (m/z > 100) of the SIMS spectra. Multivariate analyses were performed with 18 SIMS spectra (6 thicknesses x 3 areas per sample). The characteristic peaks observed for polymer chain fragmentation, $C_nH_m^+$, were observed in the fingerprint range (m/z = 0-100). The authors studied the relation between these peaks and the sample thickness and observed that their intensities are not constant as the film thickness decreases. The data analysis showed a decrease in all the additive peaks' intensities and an increase of most of the polymer peaks with the decreasing of the film thickness.

The influence of the chain length was also studied in polystyrene (additivated with Irgafos168) samples [182], with different molecular weights M_w, using multivariate analyses.

The same behaviour was observed for the PMMA and copolymer samples: decrease of the additive intensity below a threshold transition thickness. This behaviour was attributed to the segregation of the additives' molecules towards the polymer-air-interface and the flatter chains prevented the diffusion of the additives towards the free interface.

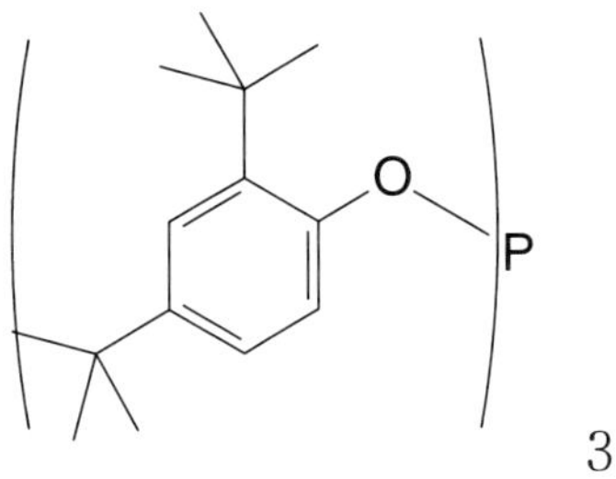

Irgafos 168

Figure 10. Chemical structure of Irgafos 168 (antioxidant additive).

MALDI-TOF MS is an efficient analytical technique to identify additives in polymers, using an internal standard approach, with the advantage of being a solvent-free method of analysis.

Schwarzinger et al. [167] performed a quantitative analysis of additives in polymers using MALDI-TOF MS. Calibration curves were constructed for the samples with different amounts of Tinuvin 770 in polypropylene, without internal standard, showing correlation coefficients between 0.911 and 0.990. These calibration curves presented a non-linear behaviour. The authors attributed this non-linear behaviour to the increasing ratio of analyte to matrix. A good fitting of the calibration curve with the experimental data would be achieved by keeping the matrix-to-analyte ratio high. MALDI-TOF calibration curves, using an internal standard, were constructed for some additives (Irganox 1014, Irganox 1010 and Irgafos 168) and presented a linear response. For Irganox 1330, no calibration curve could be obtained, probably because this molecule is hard to ionise. Summarising, this paper showed that it is possible to analyse additives without pretreatment using an internal standard and solid sample by MALDI-TOF MS.

8.3.3. Quantification of Heavy Metals in Polymers by X-ray Fluorescence Applying Calibration Methods

X-ray Fluorescence (XRF) is an efficient technique to identify metals in polymeric materials [162]. Calibration methods associated with XRF measurements can provide the contamination level of recycled polymers183].

Romão et al. [181] studied the principal contaminants of bottle-grade poly(ethylene terephthalate) (PET) samples, applying principal components (PCA) as calibration and validation method. To predict the weight percent of post-consumption poly(ethylene terephthalate) present in commercial bottle-grade PET samples, a PLS model was constructed and validated. The chemometric analysis of principal components (PCA) of the XRD spectra was used to identify the presence of metals in PET from **X** matrixes constituted by independent variables where the spectrum energies correspond to the columns and the samples correspond to the lines. The spectral region used for PCA was the region characteristic of impurities and catalyst (0-10 keV) only. To analyse the recycled fraction on PET samples, PLS was used from the ***X*** and ***Y*** matrixes, where ***X*** is the same matrix used in

PCA. The *Y* matrixes were constructed with K columns corresponding to post-consumption PET concentration in bottle-grade PET, and the rows corresponding to the samples.

A graphic of PC1 x PC2 showed separation in two groups according to the PC2 region. In the first group, the separation occurred due to the presence of the Mn catalyst (PC2 > 0). A second group was formed by five samples (PC2 < 0) where the Sb catalyst was present in all samples. While the PC1 and PC2 components showed different metals used as catalysts in the PET samples, the PC2 and PC3 scores and loadings plots showed better separation between virgin samples and samples subjected to thermo-mechanical processing.

To predict the percentage of post-consumption PET in manufactured bottle-grade PET samples, the constructed PLS model was used. A good correlation was observed for the virgin and recycled samples, where the model predicted 0 and 100 ± 3 wt% of post-consumption PET, respectively. However, for the samples containing Mn as catalyst, the prediction did not agree with the true concentration of the recycled material in the PET samples. Thus, this calibration and validation model can be applied only if the samples present the same inorganic contaminants; more sophisticated models such as PARAFAC (Parallel Factor Analysis) should be used when the samples have various elements as contaminants.

8.3.4. Simulation and Validation of Mechanical Properties of Polymers

The mechanical properties of polymers can also be simulated. Daiyan et al. [182] simulated the impact loading of injection-moulded composite of an elastomer-modified polypropylene filled with 20% of talc. The model used for predicting the mechanical properties of this composite was the linear-elastic-viscoplastic SAMP-1 model (Semi-Analytical Model for Polymers with C^1-differentiable yield surface). This model was created by ductile polymers, assuming that the polymeric specimens are homogeneous and associate pressure, yield stress and plastic dilatation in a simple damage model. Calibration was used from the uniaxial tension tests and shear and uniaxial compression tests. Two load cases were simulated: centrally loaded clamped plates and three-point bending stress. The authors discussed the calibration model in terms of the deficiencies of the calibration data, heterogeneity and anisotropy of injection moulded specimens. Anisotropy and heterogeneity of the injection moulded samples makes it difficult to predict the mechanical properties of a polymeric composite material [183]. To better evaluate the anisotropy effect in the mechanical response, the specimens were cut in parallel and perpendicular directions to the injection flow.

Experimental data and simulation using SAMP-1 as calibration method to model the impact resistance of the PP/elastomer/talc composite is shown in Figure 11. For three-point bending, the quasi-static test and impact tests at low loading rates were well predicted, but the maximum force, at high velocities (4 m/s), was underpredicted. The validation test presented different behaviours in accordance with the deflection degree. The force was underpredicted at small deflections and overpredicted at larger deflections. The authors reported that the initial underprediction could be attributed to some simplifications imposed by the model, as compressive modulus was equal to the tensile modulus in the model. The overprediction could indicate that the compressive yield stress in the model was somewhat too high. Different hardening curves in tension and compression could also be a factor. Other factors could also contribute to the overprediction, such as variation in properties through the cross-section of the bar (not included in the model) or an inadequate friction model [184].

To improve the predictive ability of the model, some aspects related to the calibration testing and the heterogeneity and anisotropy of the injection moulded specimens should be considered. Composite materials are complex materials, and in this case, it is often difficult to separate these effects, which impairs the ability to model the prediction. Using the SAMP-1 model, the authors reported that some features are missing, such as viscoelasticity. Thus, some factors should be considered, such as the cost and labour involved in the simulation versus the predictive power of the models.

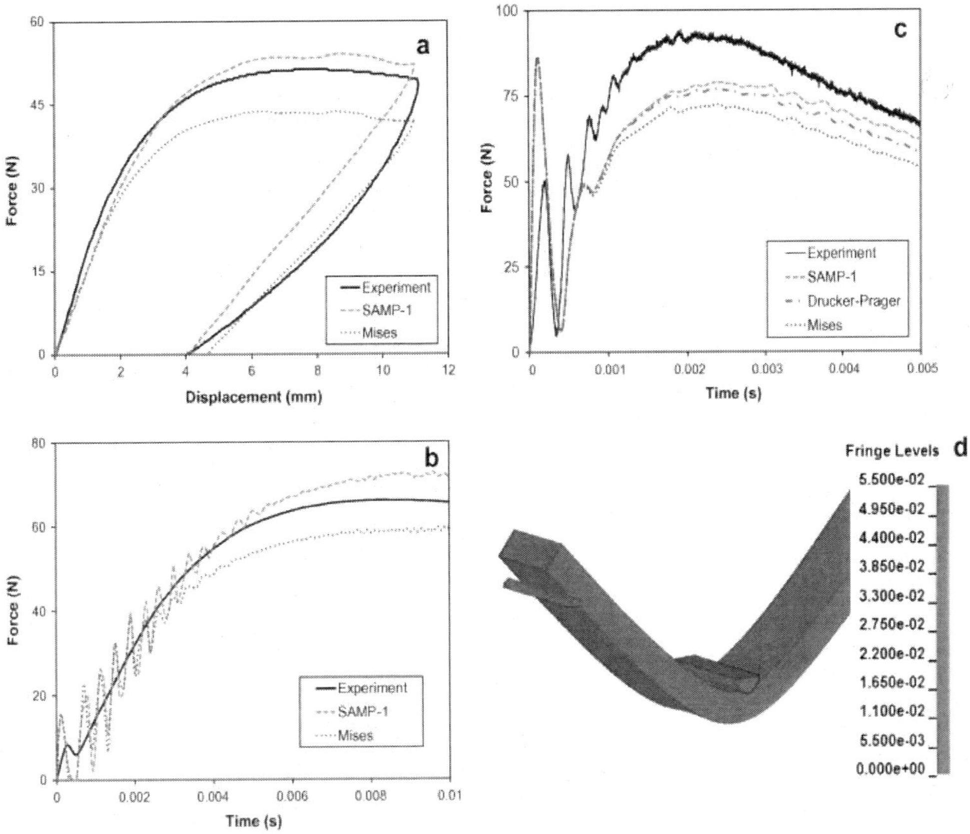

Figure 11. Mechanical properties comparing experimental and simulation results using SAMP-1 calibration model of the three-point bending test, where a) constant load and unloading rate 10 mm/min, b) 1 mm/min,c) 4 m/s, and d) plastic strain simulated after 5 ms for impact test at 4 m/s.

8.4. Conclusions and Future Perspectives of Different Calibration Methods Applied to Chemistry and Chemistry-Related Fields

In this chapter, the concepts and applications of calibration methods in chemistry were reviewed. The chapter was divided into four parts: Section 1, calibration methods in chemistry; Section 2, application of different calibration methods in pharmaceutical and inorganic analysis; Section 3, calibration methods applied to polymeric materials characterization; and Section 4, conclusions and future perspectives of different calibration methods applied to chemistry and chemistry-related fields. The first section of the chapter showed an overview of the different calibration models and the optimal conditions required for their use. In the second section, different models of univariate and multivariate calibration applied to pharmaceutical (organic) and inorganic analyses were discussed. Furthermore, different calibration concepts in chemistry were discussed, such as external calibration, standard addition, internal Standardization and isotope dilution. The third section discussed calibration methods applied to polymeric materials characterisation, which can be applied at different stages of studies involving polymers, such as in-line monitoring of extrusion process, quantitative analysis of polymer additives and quantification of heavy metals in polymers. The use of calibration methods for simulation and validation of mechanical properties of polymeric materials were also discussed.

Finally, it can be concluded that different calibration methods are being used for the quantification of different properties of interest in chemistry. Furthermore, it was noticed that the univariate calibration model based on the least squares method finds frequent application in several cases. On the other hand, a large advance has occurred in the use of multivariate methods in chemistry and chemistry-related fields, which may increasingly encourage researchers, analysts and technicians to use these methods in their activities. Currently, it can be inferred that the PLS multivariate calibration method is the most widespread one.

Thus, we can conclude that the calibration methods, both univariate and multivariate, are essential tools in laboratories, industries, research centers and institutes of metrology, as demonstrated by the large number of published studies that use calibration in chemistry.

REFERENCES

[1] K. Danzer and L. A. Currie, *Pure Appl. Chem. 70*, 993 (1998).

[2] G. M. Escandar, P. C. Damiani, H. C. Goicoechea and A. C. Olivieri, *Microchem. J. 82*, 29 (2006).

[3] E. Sanchez and B. R. Kowalski, *Anal. Chem. 58*, 496 (1986).

[4] E. Sanchez and B. R. Kowalski, *J. Chemom. 2*, 247 (1988).

[5] E. Sanchez and B. R. Kowalski, *J. Chemom. 4*, 29 (1990).

[6] I. Lavagnini and F. Magno, *Mass Spectrom. Rev. 26*, 1 (2007).

[7] D. L. Massart, B. G. M. Vandeginste, S. N. Morgan, Y. Michotte and L. Kaufman, Chemometrics: A textbook, Elsevier, Amsterdam (1988).

[8] H. O. Hartley, *Biometrika 37*, 308 (1950).

[9] B. M. G. Vandeginsten, D. L. Massart, L. M. C. Buydens, S. De Jong, P. J. Lewi. and J. Smeyers-Verbeke, *Handbook of Chemometrics and Qualimetrics: Part B*, Elsevier, Amsterdam (1998).

[10] M. Otto. *Chemometrics: Statistics and Computer Application in Analytical Chemistry*, 2nd ed., Wiley-VCH, Weihneim (2007).

[11] P. Geladi and B. R. Kowalski, *Anal. Chim. Acta 185*, 1 (1986).

[12] R. G. Brereton, *Analyst 125*, 2125 (2000).

[13] T. Naes, T. Isaksson, T. Fearn and T. Davis. *A userfriendly guide to multivariate calibration and classification.* JNIR publications, West Sussex (2001).
[14] R. Bro, *J. Chemom. 10*, 47 (1996).
[15] R. Bro, A Harshman, N. D. Sidiropoulos and M. E. Lundy, *J. Chemom. 23*, 324 (2009).
[16] M. Bahram and R. Bro, *Anal. Chim. Acta 584*, 397 (2007).
[17] H. Chen, B. Zheng and Y. Song, *J Chemom. 25*, 20 (2011).
[18] H. Chen and J. E. Kenny, *Analyst 137*, 153 (2012).
[19] L. R Tucker,. Psychometrika *31*, 279. (1966)
[20] J. D. Carroll and J. J. Chang, *Psychometrika 35*, 283 (1970).
[21] M. Kompany-Zareh, Y. Akhlaghi and R. Bro, *Anal. Chim. Acta*, *723*, 18 (2012).
[22] A. Haimovich, R. Orselli, G. M., Escandar and A. C Olivieri,. *Chemom. Intell. Lab. Syst. 80*, 99 (2006).
[23] N. M., Faber, J. Ferré, R. Boqué and J. H. Kalivas, *Chemom. Intell. Lab. Syst. 63*, 107 (2002).
[24] D. H. Shewiyo, E. Kaale, P. G. Risha, B. Dejaegher, J. Smeyers-Verbeke and Y. V. Heyden, *J. Pharm. Biom. Anal. 66*, 11 (2012
[25] T. C. O'Haver, *Anal. Chem. 51*, 91A (1979).
[26] G. L. Green and T. C. O'Haver, *Anal. Chem. 46*, 2191 (1974).
[27] A. Afkhami, M. Abbasi-Tarighat, M. Bahram and H. Abdollahi, *Anal. Chim. Acta 613*, 144 (2008).
[28] M. G. Caglayan, I. M. Palabiyik, M. Born and F. Onur, *Chem. Pap. 65*, 754 (2011).
[29] D. Özdemir and E. Dinc, *Chem. Pharm. Bull. 52*, 810 (2004).
[30] J. M. L. Gallego and J. P. Arroyo, *Anal. Bioanal. Chem. 374*, 282 (2002).
[31] N. Erk, *Il Farmaco 58*, 1209 (2003).
[32] M. I. Walash, M. S. Rizk, Z. A. Sheribah and M. M. Salin, *Chem. Cent. J. 5*, 85 (2011).
[33] M. I. Walash, M. S. Rizk, Z. A. Sheribah and M. M. Salin, *Chem. Cent. J. 5*, 85 (2011).
[34] I. A. Naguib and H. N. Darwish, *Spectrochim. Acta A 86*, 515 (2012).
[35] A. Abbaspour and R. Mirzajani, *J. Pharm. Biomed. Anal. 38*, 420 (2005).
[36] O. A. Donmez, A. Bozdogan, G. Kunt and Y. Div, *J. Anal. Chem. 65*, 30 (2010).
[37] A. S. Low and J. Wangboonskul, *Analyst 124*, 1589 (1994).
[38] M. A. Hegazy, A. M. Yehia and A. A. Mostafa, *Chromatographia 74*, 839 (2011).
[39] E. Deconinck, K. Verlinde, P. Courselle and J. O. De Beer, *J. Pharm. Biomed. Anal. 59*, 38 (2012).
[40] K. Vucicevic-Prcetic, R. Cservenak and N. Radulovic, *J. Pharm. Biomed. Anal. 56*, 736 (2011).
[41] Y. Shah, Z. Iqbal, L. Ahmad, A. Khan, M. I. Khan, S. Nazir and F. Nasir, *J. Chromat. B 879*, 557 (2011).
[42] S. Topalli, T. G. Chandrasekhar and M. M. Annapurna, *J. Chem. 9*, 832 (2012).
[43] H. M. Maher, M. A. Sultan and I. V. Olah, *Chem. Cent. 5*, 76 (2011).
[44] J. O. Grisales, J. A. Arancibia, C. B. Castells and A. C. Olivieri, *J. Chrom. B* in press. (2012).
[45] E. Dinç, A. Özdemir, H. Aksoy, O. Üstündag and D. Baleanu, *Chem. Pharm. Bull. 54*, 415 (2006).
[46] H. P. Balley, S. C. Rutan and D. R. Stoll, *J. Sep. Sci. 35*, 1837 (2012).
[47] F. Belal, A. El-Brashy, N. El-Enany and M. Tolba, *J. Fluorescence 21*, 1371 (2011).
[48] S. Bano, A. Mohd, A. A. P. Khan and K. S. Siddiqi, *J. Anal. Chem. 66*, 603 (2011).

[49] M. S. El-Din, M. Eid and A. Zeid, *Luminescence* in press, (2012).
[50] V. F. Samanidou, E. N. Evaggelopoulou and I. N. Papadoyannis, *J. Pharm. Biom. Anal.38*, 21 (2005).
[51] D. M. Fang, H. L. Wu, Y. J. Ding, L. Q. Hu, A. L. Xia and R. Q. Yu, *Talanta 70*, 58 (2006).
[52] H. W. Darwish and A. and H. Backeit. *J. Fluoresc.* in press. (2012).
[53] H. S. Shin and J. A. Oh, *J. Sep. Sci. 35*, 541 (2012).
[54] W. D'Autry, C. Zheng, K. Wolfs, S. Yarramraju, J. Hoogmartens, A. V. Schepdael and E. Adams, *J. Sep. Sci. 34*, 1299 (2011).
[55] C. Witschi and E. Doelker, *Eur. J. Pharm. Biopharm. 43*, 215 (1997).
[56] C. Caramasu, C. Madichie and R. Williams, *Trends Anal. Chem. 25*, 768 (2006).
[57] R. Otero, G. Carrera, J. F. Dulsat, J. L. Fábregas and J. Claramunt, *J. Chromatogr. A 1057*, 193 (2004).
[58] M. Khan, K. Jayasree, K. V. S. R. K. Reddy and P. K. Dubey, *J. Pharm. Biom. Anal. 58*, 27 (2012).
[59] W. Plugge and C. van der Vlies, *J. Pharm. Biomed. Anal. 14*, 891 (1996).
[60] F. W. Langkilde and A. Svantesson, *J. Pharm. Biomed. Anal. 13*: 409 (1995).
[61] S. M. Han and P. G. Faulkner, *J. Pharm. Biomed. Anal. 14*, 1681 (1996).
[62] P. Frake, D. Greenhalgh, S. M. Grierson, J. M. Hempenstall and D. R. Rudd, *Int. J. Pharm.151*, 75 (1997).
[63] N. H. Duong, P. Arratia, F. Muzzio, A. Lange, J. Timmermans and S. Reynolds, *Drug Dev. Ind. Pharm. 29*, 679 (2003).
[64] W. Plugge and C. van der Vlies, *J. Pharm. Biomed. Anal. 11*, 435 (1993).
[65] D. C. Lee and M. Webb, *Pharmaceutical Analysis*, Blackwell Publishing, Oxford (2003).
[66] M. Blanco, J. Cruz and M. Bautista, *Anal. Bioanal. Chem. 392*, 1367 (2008).
[67] C. Gendre, M. Boiret, M. Genty, P. Chaminade and J. M. Pean, *Int. J. Pharm. 421*, 237 (2011).
[68] M. A. M. Silva, M. H. Ferreira, J. W. B. Braga and M. M. Sena, *Talanta 89*, 342 (2012).
[69] T. de Beer, A. Burggraeve, M. Fonteyne, L. Saerens, J. P. Remon and C. Vervaet, *Int. J. Pharm. 417*, 32 (2011).
[70] Z. Chen, D. Lovett and J. Morris, *J. Process Contr.21*, 1467 (2011).
[71] J. D. Kirsch and J. K. Drennen, *J. Pharm. Biomed. Anal. 13*, 1273 (1995).
[72] Y. Roggo, P. Chalus, L. Maurer, C. Lema-Martinez, A. Edmond and N. Jent, *J. Pharm. Biomed. Anal. 44*, 683 (2007).
[73] M. C. Sarraguça, S. O. Soares and J. A. Lopes, *Vibr. Spectrosc. 56*, 184 (2011).
[74] M. Blanco, R. Cueva-Mestanza and A. Peguero, *Talanta 85*, 2218 (2011).
[75] M. Blanco and A. Peguero, *J. Pharm. Biomed. Anal.52*, 59 (2010).
[76] W. F. C. Rocha, A. L. Rosa, J. A. Martins and R. J. Poppi, *J. Braz. Chem. Soc. 21*, 1929 (2010).
[77] W. L. Yip, I. Gausemel, S. A. Sande and Knut Dyrstad, *J. Pharm. Biomed. Anal.70*, 202 (2012).
[78] J. A. Farrel, K. Higgins and J. H. Kalivas, *J. Pharm. Biomed. Anal. 61*, 114 (2012).
[79] N. Townshend, A. Nordon, D. Littlejohn, M. Myrick, J. Andrews and P. Dallin, *Anal. Chem. 84*, 4671 (2012).

[80] C. Eliasson, N. A. Macleod, L. C. Jayes, F. C. Clarke, S. V. Hammond, M. R. Smith, and P. Matousek, *J. Pharm. Biomed. Anal. 47*, 221 (2008).
[81] Y. Lee, J. Kim, S. Lee, Y. A. Woo and H. Chung, *Talanta 89*, 109 (2012).
[82] S. Mazurek and R. Szostak, *J. Pharm. Biomed. Anal. 40*, 1235 (2006).
[83] N. Chieng, T. Rades and J. Aaltonen, *J. Pharm. Biom. Anal. 55*, 618 (2011).
[84] R. Hilfiker, *Polymorphism in the Pharmaceutical Industry*, Wiley-VCH, Weihneim (2006).
[85] H. Uchida, T. Yoshinaga, H. Mori and M. Otsuka, *J. Pharm. Pharmacol. 62*, 1526 (2010).
[86] W. F. C. Rocha. G. P. Sabin, P. H. Março and R. J. Poppi, *Chemom. Intell. Lab. Syst. 106*, 198 (2011).
[87] C. J. Strachan, D. Pratiwi, K. C. Gordon and T. Rades, *J. Raman Spectrosc. 35*, 347 (2004).
[88] Y. Li, P. S. Chow and R. B. H. Tan, *Int. J. Pharm. 415*, 110 (2011).
[89] ISO Guide 34, *General Requirements for the Competence of Reference Material Producers*, ISO, Geneva (2009).
[90] ISO Guide 35, *Reference Materials: General and Statistical Principles for Certification*, ISO, Geneva (2006).
[91] T. Saito, T. Ihara, M. Koike, S. Kinugasa, Y. Fujimine, K. Nose and T. Hirai, *Accred. Qual. Assur. 14*, 79 (2009).
[92] G. K. Webster, I. Marsden, C. A. Pommerening and C. M. Tyrakowski, *Appl. Spectrosc. 64*, 537 (2010).
[93] S. K. Bharti and R. Roy, *Trends Anal. Chem. 35*, 5 (2012).
[94] N. H. Bings, A. Bogaerts and J. A. C. Broekaert, *Anal. Chem. 78*, 3917 (2006).
[95] A. Taylor, S. Branch, D. Halls, M. Patriarca and M. White, *J. Anal. At. Spectrom. 17*, 414, (2002).
[96] V. L. Dressler, F. G. Antes, C. M. Moreira, D. Pozebon and F. A. Duarte, *Int. J. Mass Spectrom. 307*, 149 (2011).
[97] A. P. Oliveira, R. D. Villa, K. C. P. Antunes, A. Magalhães and E. C. Silva, *Fuel 88*, 764 (2009).
[98] R. S. Amais, E. E. Garcia, M. R. Monteiro and J. A. Nóbrega, *Fuel 93*, 167 (2012).
[99] J. Sardans, F. Montes and J. Peñuelas, *Spectrochim. Acta Part B 65*, 97 (2010).
[100] J. Bettmer, M. M. Bayón, J. R. Encinar, M. L. F. Sánchez, M. R. F. Campa and A. Sanz-Medel, *J. Proteomics 72*, 989 (2009).
[101] S. R. Rocha, F. J. Sánchez-Muniz, M. Gómez-Juaristi and M. T. L. Marín, *J. Food Compos. Anal. 22*, 330 (2009).
[102] A. M. Cameán, I. Moreno, M. López-Artíguez, M. Repetto and A. G. González, *Talanta 54*, 53 (2001).
[103] K. Y. Phan-Thien, G. C. Wrigth and N. A. Lee, *Food Chem. 134*, 453 (2012).
[104] D. A. Skoog, F. J. Holler and T. A. Nieman, *Principles of Instrumental Analysis*, 5th ed., Saunders College, New York (1998).
[105] S. M. Nelms, *ICP Mass Spectrometry Handbook,* Blackwell Publishing Ltd, Boca Raton (2005).
[106] A. Montaser, *Inductively Coupled Plasma Mass Spectrometry*, 1st ed., Wiley-VCH, New York (1998).

[107] C. B. Boss and K. J. Fredeen, *Concepts, Instrumentation, and Techniques in Inductively Coupled Plasma Optical Emission Spectrometry*, 2nd ed., Perkin Elmer Corporation: Norwalk, (1997).
[108] H. E. Taylor, *Inductively Coupled plasma mass spectrometry – Practices and Techniques*, Academic Press, California (2001).
[109] R. Thomas, *Practical Guide to ICP-MS*, Marcel Dekker, Inc., New York (2004).
[110] R. S. Nascimento, R. E. S. Froes, N. O. C. Silva, R. L. P. Naveira, D. B. C. Mendes, W. B. Neto and J. B. B. Silva, *Talanta 80*, 1102 (2010).
[111] J. M. Andrade, M. J. Cal-Prieto, M. P. Goméz-Carracedo, A. Carlosena and D. Prada, *J. Anal. At. Spectrom.*, *23*, 15 (2008).
[112] J. M. Mermet, *Spectrochim. Acta, Part B 65*, 509 (2010).
[113] E. Hsieh, C. Hsiao and J. Liu, *J. Chemometrics*, *23*, 56 (2009).
[114] H. Mark and J. Workman, *Spectroscopy 20*, 26 (2005).
[115] K. G. Heumann, *Int. J. Mass Spectrom. Ion Processes 118/119*, 575, (1992).
[116] P. Rodríguez-González, J. M. Marchante-Gayón, J. I. G. Alonso and A. Sanz-Medel, *Spectrochim. Acta Part B 60*, 151 (2005).
[117] K. G. Heumann, *Anal. Bioanal. Chem. 378*, 318(2004).
[118] M. Sargent, C. Harrington and R. Harte, *Guidelines for achieving high accuracy in isotope dilution mass spectrometry (ID-MS)*, RSC Analytical methods committee, Co-coordinating Editors (2002).
[119] W. Penninckx, C. Hartmann, D. L. Massart and J. Smeyers-Verbeke, *J. Anal. At. Spectrom. 11*, 237 (1996).
[120] L. Husáková, T. Černohorský, J. Šrámková and L. Vavrušová, *Food Chem. 105*, 286 (2007).
[121] L. Cuadros-Rodríguez, L. Gámiz-Gracia, E. M. Almansa-López and J. M. Bosque-Sendra, *Trends Anal. Chem. 20*, 620 (2001).
[122] H. Emons, A. Held and F. Ulberth, *Pure Appl. Chem. 78*, 135 (2006).
[123] M. Yaman, *Anal. Biochem. 339*, 1 (2005).
[124] N. M. Caldas, S. R. Oliveira and J. A. G. Neto, *Anal. Chim. Acta 636*, 1 (2009).
[125] F. H. Lyra, M. T. W. D. Carneiro, G. P. Brandão and H. M. Pessoa and E. V. Castro *Microchem. J.*, *96*, 180 (2010).
[126] M. G. Pereira, E. R. Perreira-Filho, H. Berndt and M. A. Z. Arruda, *Spectrochim. Acta Part B 59*, 515 (2004).
[127] M. Felipe-Sotelo, J. M. Andrade, A. Carlosena and D. Prada, *Anal. Chem. 75*, 5254 (2003).
[128] A. Carlosena, P. López-Mahía, S. Muniategui, E. Fernández and D. Prada, *J. Anal. At. Spectrom. 13*, 1361 (1998).
[129] M. Grotti, M. L. Abelmoschi, F. Soggia, C. Tiberiade and R. Frache, *Spectrochim. Acta Part B 55*, 1847 (2000).
[130] M. A. Vieira, B. Welz and A. J. Curtius, *Spectrochim. Acta Part B 57*, 2057 (2002).
[131] M. Felipe-Soleto, M. J. Cal-Prieto, M. P. Gómez-Carracedo, J. M. Andrade, A. Carlosena and D. Prada, *Anal. Chim. Acta*, *571*, 315 (2006).
[132] G. A. Petrucelli, R. J. Poppi, R. L. Mincato and E. R. Pereira-Filho, *Talanta*, *71*, 620 (2007).
[133] D. C. Baxter, W. Frech and I. Berglund, *J. Anal. At. Spectrom. 6*, 109 (1991).

[134] L. A. Escudero, L. D. Martinez, J. A. Salonia and J. A. Gasquez, *Microchem. J. 95*, 164 (2010).
[135] J. Mutic, D. Manojlovic, R. Kovacevic, J. Trifunovic, N. R. Amaizah and L. Ignjatovic, *Microchem. J. 98*, 11 (2011).
[136] J. R. Souza, E. F. Santos, C. B. Duyck and T. D. Saint'Pierre, *Spectrochim. Acta Part B 66*, 356 (2011).
[137] E. S. Chaves, E. J. Santos, R. G. O. Araujo, J. V. Oliveira, V. L. A. Frescura and A. J. Curtius, *Microchem. J. 96*, 71 (2010).
[138] J. Naozuka, E. C. Vieira, A. N. Nascimento and P. V. Oliveira, *Food Chem. 124*, 1667 (2011).
[139] J. M. Mermet, *Spectrochim. Acta Part B 65*, 509 (2010).
[140] R. S. Nascimento, R. E. S. Froes, N. O. C. Silva, R. L. P. Naveira, D. B. C. Mendes, W. B. Neto and J. B. B. Silva, *Talanta 80*, 1102 (2010).
[141] M. L. Griffiths, D. Svozil, P. J. Worsfold, S. Denham and E. H. Evans, *J. Anal. At. Spectrom. 15*, 967 (2000).
[142] M. F. Pimentel, B. B. Neto, M. C. U. Araújo and C. Pasquini, *Spectrochim. Acta Part B 52*, 2151 (1997).
[143] M. Grotti, E. Magi and R. Frache, *J. Anal. At. Spectrom. 15*, 89 (2000).
[144] H. Kola, P. Perämäki and I. Välimäki, *J. Anal. At. Spectrom. 17*, 104 (2002).
[145] H. Chen, E. Dabek-Zlotorzynska, P. E. Rasmussen, N. Hassan and M. Lanouette, *Talanta 74*, 1547 (2008).
[146] K. Gerdes and K. E. Carter, *Spectrochim. Acta Part B, 66*, 712 (2011).
[147] S. R. Rocha, F. J. Sánchez-Muniz, M. Gómez-Juaristi and M. T. L. Marín, *J. Food Comp. Anal. 22*, 330 (2009).
[148] A. Ataro, R. I. McCrindle, B. M. Botha, C. M. E. McCrindle and P. P. Ndibewu, *Food Chem. 111*, 243 (2008).
[149] J. Hippler, H. W. Hoppe, F. Mosel, A. W. Rettenmeier and A. V. Hirner, *J. Chromatogr. B 877*, 2465 (2009).
[150] S. Suzuki, H. Tsuchihashi, K. Nakajima, A. Matsushita and T. Nagao, *J. Chromatogr. 437,* 322 (1988).
[151] N. Lewen, M. Schenkenberger, T. Larkin, S. Conder and H. G. Brittain, *J. Pharm. Biomed. Anal. 13*, 879 (1995).
[152] S. P. Dolan, D. A. Nortrup, P. M. Bolger and S. G. Capar, *J. Agri. Food Chem. 51*, 1307 (2003).
[153] D. Pick, M. Leiterer and J. W. Einax, *Microchem. J. 95*, 315 (2010).
[154] W. Guo, S. Hu, J. Zhang and H. Zhang, *Sci. Total. Environ. 409*, 2981 (2011).
[155] S. McSheehy, L. Yang, R. Sturgeon and Z. Mester. *Anal. Chem. 77*, 344 (2005).
[156] S. Y. Ng, A. Zou, L. P. Sim, Y. Ding, K. L. Yuen, R. Y. C. Shin and T. K. Lee, *Int. J. Mass Spectrom. 321-322*, 19 (2012).
[157] A. Donachie, A. D. Walmsley and S. J. Haswell, *Anal. Chim. Acta 378*, 235 (1999).
[158] M. Rupprecht and T. Probst, *Anal. Chim. Acta 358*, 205 (1998).
[159] L. Moberg, K. Pettersson, I. Gustavsson and B. Karlberg, *J. Anal. At. Spectrom. 14*, 1055 (1999).
[160] K. G. Lloyd, D. J. Walls and J. P. Wyre, *Surf. Interface Anal. 41*, 686 (2009).
[161] T. Rohe, W. Becker, A. Krey, H. Nägele, S. Kölle and N. Eisenreich, *J. Near Infrared Spectrosc. 6*, 325 (1998).

[162] E. T. Ozer, M. A. Cimenoglu and S. Gucer, *Instrumentation Science & Technology 39*, 357 (2011).

[163] F. M. V. Pereira, E. R. Pereira-Filho and M. I. M. S. Bueno, *J. Agric. Food. Chem. 54*, 5723 (2006).

[164] M. von Gradowski, M. Wahl, R. Forch and H. Hilgers, *Surf. Interface Anal. 36*, 1114 (2004).

[165] H. Momose, T. Maeda, K. Hattori, T. Hirano and K. Ute, *Polymer Journal 44*, 808 (2012).

[166] M. Richard-Lacroix and C. Pellerin, *Macromol. 45*, 1946 (2012).

[167] C. Schwarzinger, S. Gabriel, S. Beißmann and W. Buchberger. *J. Am. Soc. Mass Spectrom. 23*, 1120 (2012).

[168] T. Rohe, W. Becker, S. Kölle, N. Eisenreich and P. Eyerer, *Talanta 50*, 283 (1999).

[169] N. J. W. Reuvers, H. P. Huinink, H. R. Fischer and O. C. G. Adan, *Macromol. 45*, 1937 (2012).

[170] M. Yazdani-Pedram, H. Vega and R. Quijada, *Polymer 42*, 4751 (2001).

[171] W. Camacho and S. Karlsson, *J. Appl. Polym. Sci. 85*, 321 (2002).

[172] E. Şar, H. Berber, B. Aşçi and H. Cankurtaran, *Electroanalysis 20*, 1533 (2008).

[173] B. Vajna, B. Bodzay, A. Toldy, I. Farkas and T. Igricz, G. Marosi, *Express Polymer Letters 6*, 107 (2012).

[174] D. Fischer, K. Sahre, M. Abdelrehim, B. Voit, V. B. Sadhu, J. Pionteck, H. Komber and J. Hutschenreuter, *C. R. Chimie 9*, 1419 (2006).

[175] J. R. Araújo, M. R. Vallim, M. A. S. Spinacé and M-A De Paoli, *J. Appl. Polym. Sci. 110*, 1310 (2008).

[176] K. H. Esbensen, *Multivariate Data Analysis – in practice*, 5th ed., Aalborg University, Esbjerg, (2002).

[177] M-A De Paoli, *Degradação e Estabilização de Polímeros,* Ed. Artliber, São Paulo, (2009).

[178] R. Leardi, M. B. Seasholtz and R. J. Pell, *Anal. Chim. Acta 461*, 189 (2002).

[179] N. Médard and P. Bertrand, *Appl. Surf. Sci. 231-232*, 309 (2004).

[180] J. Swagten, D. Bossus and H. Vanwersch. "The calibration of XRF polyethylene reference materials with *k*0-NAA and ICP-AES", Proceedings of the 4th International k0-Users Workshop *564*, 761 (2006).

[181] W. Romão, M. F. Franco, M. I. M. S. Bueno, M. N. Eberlin and M-A De Paoli, *J. Appl. Polym. Sci. 117*, 2993 (2010).

[182] H. Daiyan, F. Grytten, E. Andreassen, H. Osnes and O.V. Lyngstad, *Materials and Design 42*, 450 (2012).

[183] M. C. Branciforti, C. A. Oliveira and J. A. Sousa, *Polym. Adv. Technol. 21*, 322 (2010).

[184] Q. Ma, X. Su, Z. Lin, J. Lasecki and X. Lai, *Polym. Compos. 31*, 587 (2010).

In: Calibration Technology, Theories and Applications ISBN: 978-1-62618-808-2
Editors: Ikumatsu Fujimoto and Kunitoshi Nishimura

Chapter 9

PRECAUTIONARY SAVINGS AND WEALTH INEQUALITY: A GLOBAL SENSITIVITY ANALYSIS IN A CALIBRATED MACROECONOMIC MODEL

Marco Cozzi*
Department of Economics, Queen's University, Ontario, Canada

ABSTRACT

This chapter applies a Monte Carlo calibration methodology in a macroeconomic model with labor income risk, incomplete markets, and heterogeneous agents. This is a version of the benchmark Aiyagari 1994 economy, a general equilibrium model designed to quantify the size of precautionary savings and the degree of wealth inequality. The calibration procedure assumes parameter uncertainty, implemented in the quantitative analysis with empirically motivated prior distributions for the parameters. The main objective is to conduct a global sensitivity analysis, and several computational experiments are performed, allowing for the priors to differ in both their shape and dispersion. Although the results of this global robustness analysis are broadly consistent with Aiyagari's original findings, they highlight some novel issues. Even when considering priors for the parameter uncertainty that are highly dispersed, leading to several extreme calibrations, the size of precautionary savings found in the sampling distributions is always modest: at most, they account for an 11% increase in the saving rate. However, it is shown that the parameter representing the exogenous borrowing limit leads to relatively large changes in measures of wealth inequality. The wealth Gini index increases by 15 points when considering values of the borrowing limits that lead to empirically plausible shares of households with a negative net worth. Moreover, the sampling distributions of the equilibrium outcomes allow studying which parameters have the greatest influence. With an appropriate regression analysis, it is found that the parameters that have the largest effects on the Gini index are the capital share, the borrowing limit, and the depreciation rate. Similarly, the parameters affecting most significantly the precautionary savings are the risk aversion and the standard deviation of

* Corresponding Author's address: Dunning Hall, 94 University Avenue, Kingston, ON K7L 3N6, Canada. Email: mcozzi@econ.queensu.ca.

the income shocks. These findings suggest that these parameters should be carefully calibrated in similar environments, because the quantitative results are likely to be deeply affected by their chosen values.

Keywords: Calibration, Precautionary Savings, Wealth Inequality, Heterogeneous Agents, Monte Carlo

9.1. Introduction

The *calibration* methodology is relatively recent in the field of economics. It was introduced as a macroeconometric tool by Kydland and Prescott 1982, the first contribution explicitly proposing this empirical approach to pin down the values of the free parameters included in a theoretical model. Since then, a still on-going paradigm shift in macroeconomics accompanied by both an unprecedented increase in computational power and a wider diffusion of computational methods for economists, has lead this empirical tool to become one of the most popular ways to address quantitative questions in macroeconomics.

Over the years, the way research in macroeconomics is performed has substantially changed. The field has moved away from the prevalence of simple theoretical models that privileged analytical tractability over realism. The revolution lead by Kydland, Lucas, Prescott, Sargent, Sims, and many others was motivated by: 1) a perceived fracture between the theories that were being developed and their empirical predictions (Prescott 1986), and 2) the impossibility to study many issues that are of greatest importance to macroeconomists, such as the distributional impacts of a policy change (Hansen and Heckman 1996).

The philosophy underlying the *calibration* movement has been imported in economics from the physical sciences. Kydland and Prescott 1996, a contribution that can be considered the manifesto of computational experiments in economics, describe calibration as follows:

> "Originally, in the physical sciences, calibration referred to the graduation of measuring instruments. [...] In a sense, model economies, like thermometers, are measuring devices. Generally, some economic questions have known answers, and the model should give an approximately correct answer to them if we are to have any confidence in the answer given to the question with unknown answer. Thus, data are used to calibrate the model economy so that it mimics the world as closely as possible along a limited, but clearly specified, number of dimensions."

Since its infancy in the late 80's and early 90's, the calibration method in macroeconomics has evolved. Currently, there are three main approaches to empirical macroeconomics. The first approach, as stated by Prescott 1986 and Kydland and Prescott 1996, rejects any formal statistical procedure to match some dimensions of the data, arguing that any macroeconomic model is too abstract, hence necessarily false. However, Bayesian econometricians have stressed that this calibration method can be interpreted as an extreme case of Bayesian analysis, which specifies a degenerate prior for the calibrated parameters. The second approach embraces more formal statistical procedures to select parameter values, and to perform hypothesis testing, with the Simulated Method of Moments and Indirect Inference being the two most common estimation methods. The third one is the Bayesian analysis, where parameter priors and the likelihood function are exploited to compute the posterior densities for both the parameters and the outcomes of interest.

More in detail, the typical calibration procedure envisioned by Kydland and Prescott consists of a series of steps.[1] The first step is related to the specification of an economic model, tailored to address the question under investigation, say the determinants of saving behavior. The theoretical model specifies an economic environment where a set of agents participate in several markets, for example by purchasing consumption goods, selling labor services, and lending or borrowing resources in financial markets. In the neoclassical tradition, agents behave rationally, and make their choices in order to maximize a well-posed criterion, the utility function, which allows them to rank different decisions in an uncertain environment. By obtaining the agents' decision rules, and aggregating them appropriately, the model delivers some predictions on a set of endogenous variables, say the interest rate and the share of output paid to workers. A calibration method is needed, because the model usually does not possess an analytical solution: it must be solved numerically, and simulated with appropriate techniques. It goes without saying that the quantitative results will depend on the specific values used for the parameters. Some guidance is needed in the "quest" for an appropriate choice for the parameters values, with the data helping the researcher in the calibration of the model parameters. A set of observables are needed, which directly relate to the model's endogenous variables. Having gathered the data $\mathbf{x}_j$ and having computed, say, the average interest rate or the average labor share, that is a set of sample statistics $h(.)$ stacked in the n×1 vector $\mathbf{z} = h(\mathbf{x}_j)$, the calibration method will search in the parameters space for the model to achieve a good fit. More formally, denote with $\boldsymbol{\theta}$ the m×1 vector of parameters included in the model, and with $\mathbf{z}^{\mathbf{M}}(\boldsymbol{\theta})$ the n×1 vector stacking the model counterpart of the same statistics computed in the data. In general, the typical calibration method will choose the set of parameters $\widehat{\boldsymbol{\theta}}$ such that:

$$\widehat{\boldsymbol{\theta}} = \underset{\boldsymbol{\theta}}{\mathbf{Argmin}}\, G(\mathbf{z}^{\mathbf{M}}(\boldsymbol{\theta}) - \mathbf{z})$$

In order to make this criterion operational, a researcher needs more information. First and foremost, a specific loss function $G(.)$ is needed. Unfortunately, there is no consensus on which loss function should be used. Obviously, different choices lead to different approaches: this is of particular importance whenever $\mathrm{m} < \mathrm{n}$, namely when the number of parameters is less than the number of statistics computed in the data. The most common choice sets $\mathrm{m} = \mathrm{n}$, i.e. it uses as many statistics as model parameters, and relies (with some abuse of language) on an exactly identified model, namely a model that just adds up the errors made in all the moments, i.e. it uses $G(.) = \sum_{s=1}^{m} \left| \mathbf{z}_{\mathbf{s}}^{\mathbf{M}}(\boldsymbol{\theta}) - \mathbf{z}_{\mathbf{s}} \right|$.

A few more comments are in order. At this stage, this calibration methodology *cannot* be interpreted as an implementation of the statistical device known as the method of moments. In the latter, it is assumed that there is one (and only one) vector of parameters that has generated the data. As a consequence, all the moments need to be true in a statistical sense, otherwise the model would be rejected by the data. Differently, the standard calibration methodology does not test formally if some of the moments are violated, aiming for a less

[1] Lower-case bold variables stand for vectors, the variable j denotes the unit of observation, and $\boldsymbol{h}(.)$ denotes a known vector-valued function, collecting a set of averages, variances or correlations.

ambitious goal: achieving the best possible fit. Moreover, notice that the data requirements are less binding than for other empirical methods. It is not required that all the variables in the dataset come from the same units of observation (as, for example, in a maximum likelihood estimation procedure). It is therefore possible to use different data sources, with each dataset contributing different observables, as long as the sampling scheme is coherent across datasets, and they refer to the same underlying population. This is an important feature, because it allows studying problems for which there is no ideal dataset containing all the information needed for the analysis.

9.1.1. A Non-standard Calibration Approach, with an Application to Savings and Wealth Inequality

In this chapter, I will consider a benchmark macroeconomic model with heterogeneous agents. This model will be calibrated with a Monte Carlo methodology, which embeds a global sensitivity analysis. This methodology can be interpreted as a mix of the first and the third approaches outlined above. The model will not be formally tested, but the calibration will include a distinctive element of the Bayesian analysis: the prior distributions for the parameters (hence the sampling distributions of the model outcomes).

The model is designed to deal with the households' saving behavior, and an endogenously determined wealth distribution. It represents a simple (yet powerful and easily extendable) framework, which has become one of the backbones to conduct quantitative research in macroeconomics. In particular, it tackles the following questions: What is the size of precautionary savings? What accounts for the degree of wealth inequality observed in the data? These questions are hard to address relying solely on data regarding income, asset levels and consumption. A quantitative theory is a useful approach to understand how income uncertainty and limited insurance opportunities affect the aggregate savings and the shape of the wealth distribution.

After the seminal paper by Aiyagari 1994, several contributions have tried to measure the size of precautionary savings, typically finding that they are small, and that the standard incomplete markets model cannot account for the high degree of wealth concentration found in the data.[2]

The limited scope of the robustness checks that are usually carried out in quantitative macroeconomics analyses stands out as a possible source of criticism on the results obtained in this literature. In order to tackle this kind of objections, in this chapter I apply Canova 1994 methodology to perform a thorough sensitivity analysis for the Aiyagari 1994 economy, making two contributions. On the one hand, this procedure confirms Aiyagari 1994's main findings, which suggest that the size of precautionary savings is small and that uninsurable income risk is not enough to explain the observed wealth inequality. On the other hand, it shows that with today's computational power global sensitivity analysis of the type proposed by Canova 1994 in the simple Real Business Cycle framework are also feasible for richer Heterogeneous-Agent (*HA*) economies.

[2] See, for example, the surveys by Cagetti and De Nardi 2008, Heathcote, Storesletten and Violante 2009, and Quadrini and Rios-Rull 1997.

This type of exercise is important for two reasons. First, often there is no natural empirical counterpart to the model's parameters, hence some relevant calibration targets (e.g. the capital/output ratio, or the labor share) have to be chosen for the model to match. In the absence of aggregate uncertainty, the researcher is confronted with two possible choices. The first one focuses the attention on a single year, making the measurements for all variables consistent with the steady-state of the model, whereas the second choice takes long-run averages of the statistics of interest instead. Moreover, selecting the moments relevant for the analysis together with their weighting matrix can be a delicate step. Either way, sampling variability and parameter uncertainty for the parameters that are taken from related empirical studies (such as the ones characterizing the exogenous stochastic processes for labor income, or the degree of risk aversion) can cast doubts on the generality of the quantitative findings. Second, a carefully executed global robustness analysis can deliver bounds and distributions for the measurement of the economic outcomes under study, showing how likely these outcomes are. A more widespread implementation of this methodology can give the calibration approach to empirical research sounder evidence for its findings.

The results of this global robustness analysis are consistent with Aiyagari's findings. Even when considering priors for the parameter uncertainty which are very different from each other (Uniform Vs. Symmetric and Asymmetric Beta distributions) and highly dispersed, the size of the precautionary savings is modest and the degree of wealth concentration falls short of the empirical observations. However, the results show that the parameter representing the exogenous borrowing limit seems to lead to relatively large changes in measures of wealth inequality. The (median) wealth Gini index increases by more than 10 points when considering values of the borrowing limits that lead to empirically plausible shares of households with a negative net worth, moving from 0.45 to up to 0.61. This number, however, is still well below its empirical counterpart of 0.80, as found for the U.S. by Budria Rodriguez, Diaz-Gimenez, Quadrini, and Rios-Rull 2002.

Finally, the simulated data make it possible to study which parameters affect the outcomes most significantly. As for the wealth Gini index, the parameters that quantitatively have the largest effects are the capital share, the borrowing limit, and the depreciation rate. As for the precautionary savings, the parameters affecting them the most are the risk aversion and the standard deviation of the income shocks.

The rest of the chapter is organized as follows. Section 9.2 presents the theoretical model. Section 9.3 is devoted to the description of the Monte Carlo experiments. Section 9.4 presents the main results, while Section 9.5 concludes. A series of appendices provide more details on the methodology together with some additional results.

9.2. THE AIYAGARI 1994 ECONOMY

This is a General Equilibrium (*GE*) model with incomplete markets and uninsurable income risk, designed to quantify the size of precautionary savings and the degree of wealth inequality.[3] Time is discrete. The economy is populated by a measure one of infinitely lived ex-ante identical agents.

[3] In the interest of space, just a sketch of the model is presented. For more details see Appendix A, Aiyagari 1994 and Rios-Rull 1999.

9.2.1. Preferences

Agents' preferences are assumed to be represented by a time separable utility function U(.). Agents' utility is defined over stochastic consumption sequences $\{c_t\}_{t=0}^{\infty}$: their aim is to choose how much to consume (c_t), and how much to save in an interest bearing asset (a_{t+1}) in each period of their lives, in order to maximize their objective function. The agents' problem can be defined as follows:

$$\max_{\{c_t,a_{t+1}\}_{t=0}^{\infty}} E_0 U(c_0, c_1, \ldots) = \max_{\{c_t,a_{t+1}\}_{t=0}^{\infty}} E_0 \sum_{t=0}^{\infty} \beta^t u(c_t)$$

where E_0 represents the expectation operator over the stochastic efficiency units of labor ε_t, and $\beta \in (0,1)$ is the subjective discount factor. I assume that $u(c_t) = \frac{c_t^{1-\sigma}-1}{1-\sigma}$: the per period utility function is strictly increasing, strictly concave, satisfies the Inada conditions, and has a Constant Relative Risk Aversion equal to σ.

9.2.2. Endowments

All agents are born with the same asset endowment a_0. There is a stochastic process for the effective units of labor (ε) that a worker is going to supply in the labor market. This process is assumed to be an exogenous continuous first order Markov process, specified as an AR(1), with ρ_y being the persistence parameter, and σ_y^2 the variance of the innovations: $\varepsilon_{t+1} = \rho_y \varepsilon_t + \eta_{t+1}, \eta_t \sim iid\ N(0, \sigma_y^2)$.

9.2.3. Production

The production side of the model is represented by a constant returns to scale technology of the Cobb-Douglas form, which relies on aggregate capital K_t and labor L_t to produce the final output Y_t: $Y_t = F(K_t, L_t) = K_t^{\alpha} L_t^{1-\alpha}$, where α stands for the capital share of output. The labor input L_t is the sum of the workers' efficiency units $L_t = \int \varepsilon_t d\mu_{\varepsilon,t}$, where $\mu_{\varepsilon,t}$ is the distribution over the labor endowments implied by the markov process.

9.2.4. Other Market Arrangements

All markets are competitive. Capital is supplied by rental firms that borrow from workers at the risk-free rate r and invest in physical capital, which depreciates at rate δ.

There are no state-contingent markets to insure against income risk, but workers can self-insure by saving into the risk-free asset. The agents also face a borrowing limit, denoted as $b \geq 0$.

Table 1. Model parameters and their support

Parameter	*Description*	*Minimum*	*Maximum*
Time Period	*Yearly*		
α	*Capital share*	$\underline{\alpha}$=0.25	$\overline{\alpha}$=0.45
β	*Rate of time preference*	$\underline{\beta}$=0.94	$\overline{\beta}$=0.98
δ	*Capital depreciation rate*	$\underline{\delta}$=0.03	$\overline{\delta}$=0.12
σ	*CRRA*	$\underline{\sigma}$=0.5	$\overline{\sigma}$=3.5
ρ_y	*Persistence of the earnings process*	$\underline{\rho_y}$=0.8	$\overline{\rho_y}$=0.99
σ_y	*S.d. of the earnings shocks*	$\underline{\sigma_y}$=0.04	$\overline{\sigma_y}$=0.25
b	*Borrowing limit*	$\underline{b}$={0,-1,-2,-2.5}	$\overline{b}$=0

9.3. Aiyagari QJE 1994 Meet Canova JAE 1994

Can we complement standard calibration methods by undertaking more robust (i.e. global) sensitivity analysis in equilibrium HA models? The answer is yes, by applying the methodology proposed by Canova 1994, in order to parameterize the model economy.

Table 1 provides a list of the parameters in the Aiyagari 1994 economy, together with their upper and lower values that are going to be used in the simulations. The economy under study has seven independent parameters: α, β, δ, σ, ρ_y, σ_y and b. Canova 1994 provided a very simple procedure to implement global robustness checks for fully parameterized quantitative macroeconomic models, which has not been fully exploited by the applied general equilibrium calibration community. The essential feature of Canova's approach is to acknowledge parameter uncertainty. This uncertainty can come from uncertainty on the moments to be matched, or from the intrinsic uncertainty of the estimation studies whose results are used to pin down exogenously some parameters. Rather than computing the quantitative implications of a theory for a unique set of calibrated parameters, the methodology assumes prior distributions for them, and solves repeatedly the model for many, many different calibrations obtained by drawing each parameter vector from the prior distributions.

Notice that this type of exercise is not meant to be a substitute for more traditional calibration studies. Once a quantitative theory is proposed and a set of empirically relevant values for the parameters are available, one can provide more robust evidence on the original quantitative findings, possibly suggesting in which cases and, more importantly, for which economic channels some results are more likely to be observed.

9.3.1. Monte Carlo Calibration as a Global Sensitivity Analysis

The calibration procedure consists of performing a series of Monte Carlo experiments, which rely on two steps.

In the first step, the prior distributions from which the parameters are going to be drawn are postulated. For the application in this study, I will specify priors that are distributed as Uniform and Beta random variables.

In the second step, 2,500 economies are simulated and solved. These economies differ only in the parameter vector that is drawn at each iteration (*i*). For every economy, two equilibria are computed: the equilibrium of the Incomplete Markets economy (*IM*), and the equilibrium of the corresponding Complete Markets one (*CM*). Notice that the computation of the equilibrium in the *IM* economy requires an iterative procedure on the interest rate, until the asset market clearing is achieved. Differently, the equilibrium in the *CM* economy is easily found by exploiting the condition $1 = (1 + r)\beta_i$. In the *CM* economy, the equilibrium interest rate is pinned down by the discount factor: it will depend on the actual realization of this parameter in each Monte Carlo replication. Every iteration *i* will deal with a different β_i, hence it will involve a different equilibrium interest rate $r_i = (1/\beta_i) - 1$, and different allocations. Substituting the equilibrium r_i into the firms' first order conditions gets the aggregate capital in the *CM* economy for each simulated combination of parameters (which in turn is going to be affected by the draws of α_i and δ_i, determining the marginal productivity and the user cost of capital, and by the draws of $\rho_{y,i}$ and $\sigma_{y,i}$, determining the total labor input L_i).

Finally, the size of the precautionary savings is quantified as the difference between the aggregate savings in the *IM* economy and the aggregate savings in the *CM* one.

These experiments are repeated several times: with different prior distributions for the parameters, and, as an additional robustness check, with different assumptions on the degree of market incompleteness (i.e. on the tightness of the borrowing constraint).[4]

Table 2. Experiments - Uniform and Beta prior distributions over a generic parameter p

Experiment	*Priors*	*B. Limit*		
Uniform 1 (*U1*)	$U(\underline{p},\overline{p})$	$\underline{b}=\overline{b}=0$		
Uniform 2 (*U2*)	$U(\underline{p},\overline{p})$	$\underline{b}=-1.0$		
Uniform 3 (*U3*)	$U(\underline{p},\overline{p})$	$\underline{b}=-2.0$		
Uniform 4 (*U4*)	$U(\underline{p},\overline{p})$	$\underline{b}=-2.5$		
Beta 1 (*B1*)	$Beta(2,2;\underline{p},\overline{p})$	$\underline{b}=-2.5$		
Beta 2 (*B2*)	$Beta(10,10;\underline{p},\overline{p})$	$\underline{b}=-2.5$		
Beta 3 (*B3*)	$Beta(5,2;\underline{p},\overline{p})$	$\underline{b}=-2.5$		
Beta 4 (*B4*)	$Beta(2,5;\underline{p},\overline{p})$	$\underline{b}=-2.5$		
			Prior for ρ_y:	***Prior for*** σ_y:
Beta 5 (*B5*)	$Beta(10,10;\underline{p},\overline{p})$	$\underline{b}=-2.5$	$Beta(10,10;\underline{\rho_y},\overline{\rho_y})$	$Beta(5,2;\underline{\sigma_y},\overline{\sigma_y})$
Beta 6 (*B6*)	$Beta(10,10;\underline{p},\overline{p})$	$\underline{b}=-2.5$	$Beta(5,2;\underline{\rho_y},\overline{\rho_y})$	$Beta(10,10;\underline{\sigma_y},\overline{\sigma_y})$
Beta 7 (*B7*)	$Beta(10,10;\underline{p},\overline{p})$	$\underline{b}=-2.5$	$Beta(5,2;\underline{\rho_y},\overline{\rho_y})$	$Beta(5,2;\underline{\sigma_y},\overline{\sigma_y})$

[4] In order to minimize the effect of sampling variability, for the same values of $\rho_{y,i}$ and $\sigma_{y,i}$, all simulations share the same sequence of income shocks. The solution algorithm, some computational aspects and the simulation details are explained in Appendices B and C.

Table 2 reports the list of prior distributions that are going to be used. More in detail, independent Uniform and Beta priors are specified.

The choice settled on these specific distributions for two reasons: 1) with the priors centered around the typical calibration values, it is possible to assess if there is a tendency for the model to "compress" the outcomes around a set of results; 2) by comparing the results of the Uniform and Beta calibrations, it is possible to appreciate the effect of relying on symmetric vs. asymmetric priors, which, at the same time, differ in their dispersion.

The first set of experiments, denoted by *U1-U4*, assumes that all parameters are uniformly distributed. The average of such distributions corresponds to the values that are used in the typical calibrations of these models: $\alpha^{avg} = 0.35$, $\beta^{avg} = 0.96$, $\delta^{avg} = 0.075$, $\sigma^{avg} = 2$, $\rho_y^{avg} = 0.9$, and $\sigma_y^{avg} = 0.145$.[5] These values are chosen to match some macroeconomic facts and to be consistent with the available empirical evidence. The first three parameters (α^{avg}, β^{avg} and δ^{avg}) match the capital share of income (which is approximately 35%), the capital/output ratio (approximately 3), and the investment/output ratio (approximately 20%), while the remaining three are borrowed from related applied studies.[6] In order to check if the findings are robust to relatively large changes in the parameters, and if they are well behaved with respect to them, the parameters' bounds are chosen to allow for a wide range of possible calibrations. At the same time, an attempt is made to prevent the model to grossly miss some quantitative implications. This would happen in several replications, if the parameters' ranges were to be too wide.

The four uniform experiments differ in only one aspect, which is related to the borrowing constraint b. For the Monte Carlo simulations, case *U1* is going to assume that agents cannot borrow, namely an economy for which the parameter b is fixed at its most stringent value, $b = \overline{b} = \underline{b} = 0$. This is the case that is most often considered in the literature, including Aiyagari (1994). The other three Uniform prior cases, *U2-U4*, relax this aspect, allowing for borrowing constraints that vary from one experiment to the other (while keeping the sequence of realizations for all the remaining parameters the same in all replications). These additional cases rely on a uniform prior, whose upper bound is the no borrowing case (i.e. $\overline{b} = 0$), while the lower bound is less and less stringent, being equal to $\underline{b} = \{-1.0, -2.0, -2.5\}$. These three values were chosen because on average they imply agents with a negative net worth equal to 8%, 11% and 15%, respectively. These values are consistent with the data reported in Cagetti and De Nardi (2008) and Wolff (1998).

Table 2 also reports the second set of prior distributions. In these experiments, denoted by *B1-B7*, more flexibility is allowed for. It is assumed that all the parameters are distributed according to a Beta distribution. This distribution has four parameters: two shape parameters together with two location parameters. The location parameters allow to define the support of the distribution in an interval different from the [0,1] one, and are kept constant across all experiments. For comparability with the Uniform cases, these coincide with the upper and

[5] The length of the model period is set to one year. I also considered another specification, with $\rho_y = 0.5$, which is not reported. The results related to the precautionary savings are very similar, while the ones related to inequality are slightly different. However, the empirical studies on income risk provide evidence dismissing values of the persistence parameter below 0.8.

[6] See, among others, Attanasio, Banks, Meghir, and Weber 1999, Guvenen 2009, and the papers cited therein.

lower bounds reported in the last two columns of Table 1.[7] As for the other two (shape) parameters that are needed to fully characterize the Beta distribution, two different set-ups are considered. Four cases, *B1-B4*, rely on the priors Beta(2,2), Beta(10,10), Beta(5,2), Beta(2,5) for all the parameters. The first two specifications deal with a Beta distribution which is symmetric around its mean, because the two shape parameters are equal. The pair (10,10) implies a distribution which is substantially more concentrated around its mean, when compared to the (2,2) one. The other two cases deal with asymmetric priors. The Beta(5,2) is a left skewed distribution, while the Beta(2,5) is a right skewed one. Figure (1) shows the shape of the four Beta priors over the [0,1] interval.

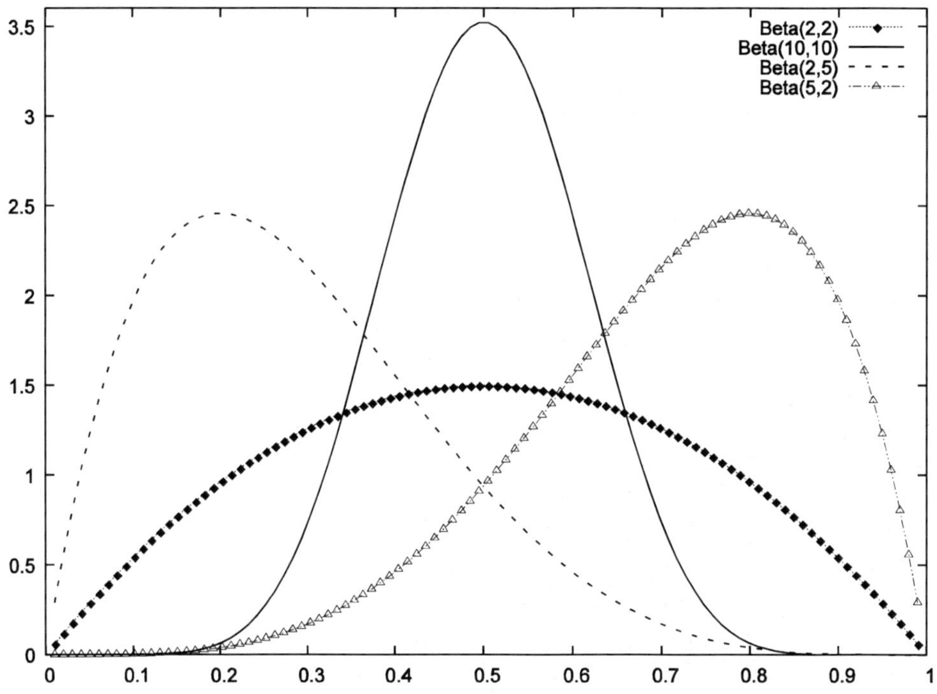

Figure 1. Monte Carlo Experiments, Beta(p,q) priors.

The last three cases, *B5-B7*, introduce an empirically motivated mix of Beta priors. All parameters, but ρ_y and σ_y, rely on the Beta(10,10) prior, i.e. the most concentrated case. As it will be discussed below, this assumption was found to minimize the number of implausible quantitative implications of the model. However, for the two parameters describing the earnings process, I rely on either a Beta(5,2), or a Beta(10,10) prior. The reason is simple: the bulk of the available econometric evidence, discussed for example in Guvenen 2009, tends to find highly persistent processes, with many studies estimating $\rho_y > 0.92$. However, depending on the specification of the transitory component of the income shocks and whether income profile heterogeneity is included in the econometric model, this estimate can be as low as 0.8. These results suggest a reasonable range for the earnings persistence parameter,

[7] Notice that I kept exactly the same bounds for all parameters as in the uniform case that was matching the average share of households in debt across simulations found by Wolff 1998, that is case *U4*, with $\underline{b} = -2.5$.

and that one should try to analyze several priors that give different importance to some subset of the parameter's support. Similar comments apply to σ_y, the standard deviation of the innovations.

Table 3. Monte Carlo Results – Precautionary Savings Statistics (% Change)

Case	*Min*	*Max*	*Mean*	*Median*	*S.d.*
U1	.001	11.0	1.3	.8	.01
U2	.001	10.7	1.1	.7	.01
U3	.001	10.7	1.1	.7	.01
U4	.001	10.7	1.1	.6	.01
B1	.001	6.8	1.0	.8	.01
B2	.001	2.9	.9	.9	.01
B3	.001	5.3	1.5	1.4	.01
B4	.001	3.2	.4	.3	.01
B5	.001	4.2	1.7	1.6	.01
B6	.001	2.5	.7	.7	.01
B7	.001	4.3	1.3	1.2	.01

Table 4. Monte Carlo Results - Wealth Gini Index, (restricted cases in parenthesis with the exception of case U1, because one restriction does not apply)

Case	*Min*	*Max*	*Mean*	*Median*	*S.d.*
U1	.24	.55	.45	.45	.03
	(-)	(-)	(-)	(-)	(-)
U2	.25	.76	.50	.49	.06
	(.41)	(.61)	(.52)	(.52)	(.04)
U3	.25	.85	.54	.53	.09
	(.44)	(.61)	(.53)	(.53)	(.04)
U4	.25	.88	.56	.54	.09
	(.42)	(.61)	(.54)	(.54)	(.04)
B1	.36	.82	.56	.55	.07
	(.44)	(.61)	(.54)	(.54)	(.04)
B2	.46	.71	.56	.56	.03
	(.49)	(.61)	(.56)	(.56)	(.02)
B3	.39	.72	.51	.50	.04
	(.46)	(.60)	(.53)	(.52)	(.03)
B4	.43	.84	.61	.61	.06
	(.47)	(.61)	(.56)	(.57)	(.03)
B5	.46	.72	.57	.57	.04
	(.47)	(.61)	(.56)	(.56)	(.02)
B6	.48	.69	.57	.57	.03
	(.49)	(.61)	(.56)	(.56)	(.02)
B7	.47	.71	.57	.57	.04
	(.48)	(.60)	(.56)	(.57)	(.02)

9.4. Results

This Section presents the main results. First, I show that allowing for very flexible calibrations does not alter considerably typical measures of precautionary savings. Second, I discuss how measures of wealth inequality behave in this global robustness exercise.

Tables 3 and 4 report a set of statistics for the two outcomes that are the focus of my analysis. Table 3 deals with the percentage change in aggregate savings, while Table 4 with wealth inequality. The minimum, maximum, mean, median and the standard deviation of these variables are listed for all priors.

Moreover, in order to provide a more intuitive summary of such a rich analysis, several figures are included. All figures show eleven panels. Each panel represents the non-parametric kernel density estimate of a variable for a particular prior. The first four panels deal with the Uniform experiments, while the remaining seven deal with the Beta ones.

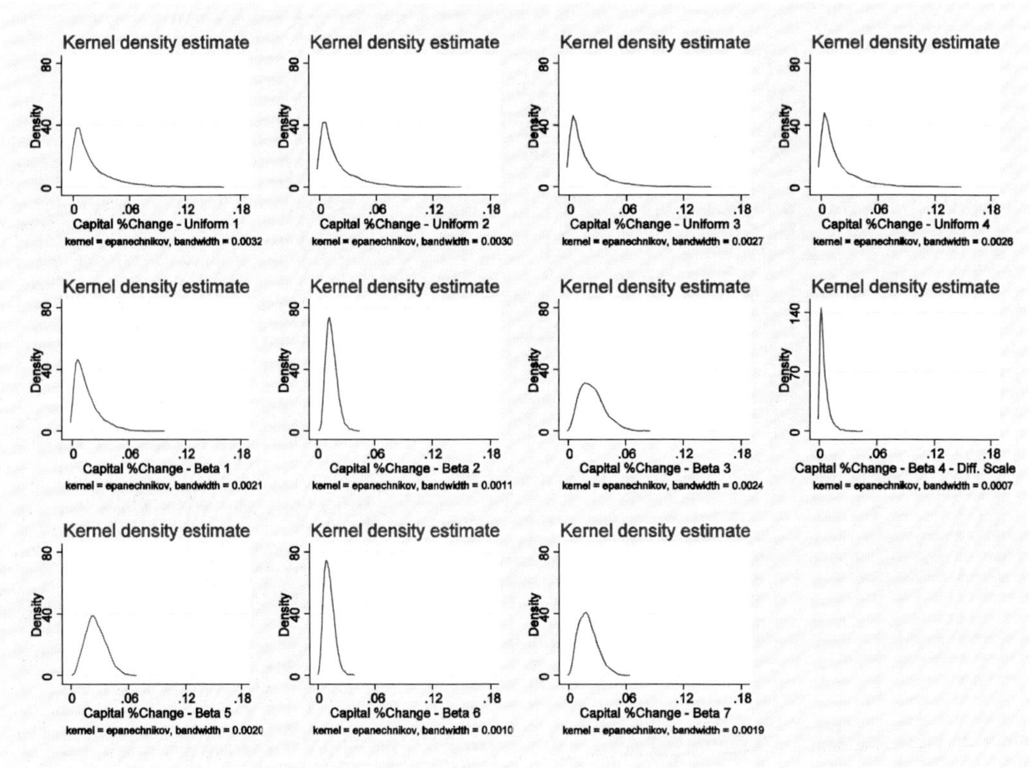

Figure 2. Precautionary Savings - Estimated Densities for the eleven Monte Carlo Experiments.

9.4.1. Precautionary Savings

Figure (2) displays a measure of precautionary savings, the percentage increase in the aggregate saving rate of the *IM* economy compared to the *CM* one. The results are striking. When considering a set of eleven Monte Carlo calibrations of size 2,500, each coming from a different prior, the percentage increase in the aggregate saving rate is always modest. There are virtually no cases in which the increase is as high as 6%, with the bulk of the replications being included in the 0.5%-3% increase range.

In Table 3 we see that the maximum increase in the saving rate is 11%, which is found in experiment *U1* (not surprisingly, the no borrowing case), while the minimum increase is always very close to zero. In eight out of the eleven priors, the median increase is less than a percentage point. According to this quantitative theory, precautionary savings are very small indeed.

These results could be potentially driven by either an odd behavior of the saving rate, or by the model failing to account for some other features of the data related to the saving behavior, such as the capital/output ratio. As Figures (3) and (4) display, generally this is not the case. The saving rate falls in reasonable ranges, as the capital/output ratio does. There are indeed some replications that imply values that are grossly over or understated, just like the extreme cases in panels 1 to 4, namely the Uniform priors. In order to tackle this potential issue, I provide some additional robustness checks below. However, when focusing on the Beta cases, on *B2* in particular (which has the least counterfactual quantitative implications), we see that precautionary savings are highly concentrated around the 1% value.[8]

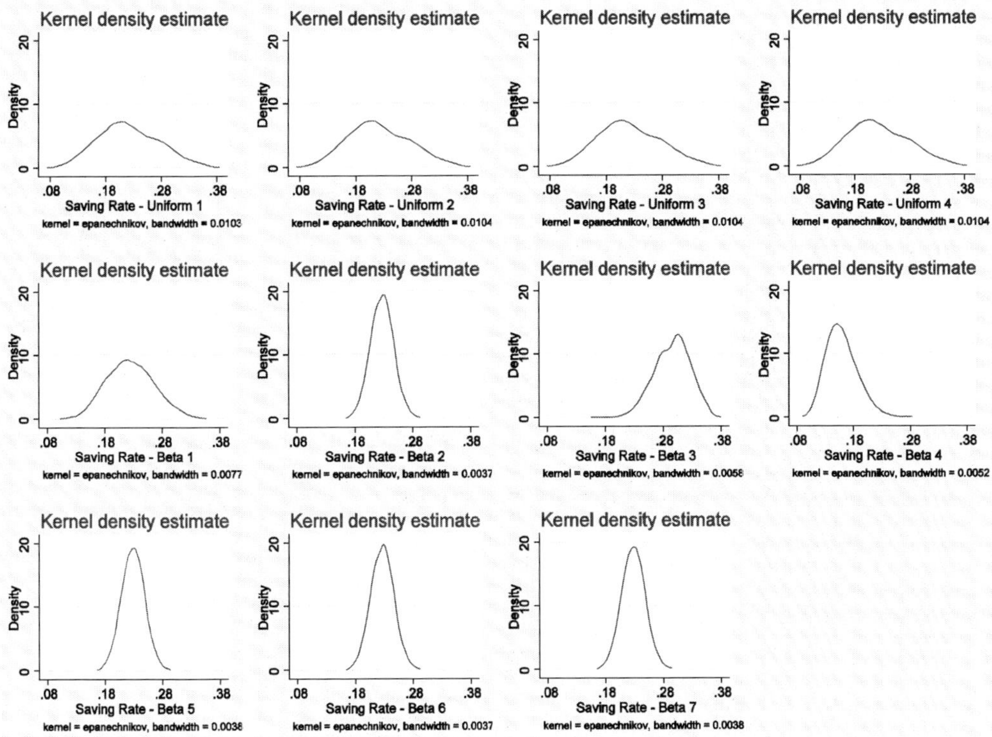

Figure 3. Saving Rates - Estimated Densities for the eleven Monte Carlo Experiments.

[8] This very set of comments applies also when considering another measure of precautionary savings, the percentage increase in the economy's capital stock, when moving from the CM economy to the IM one.

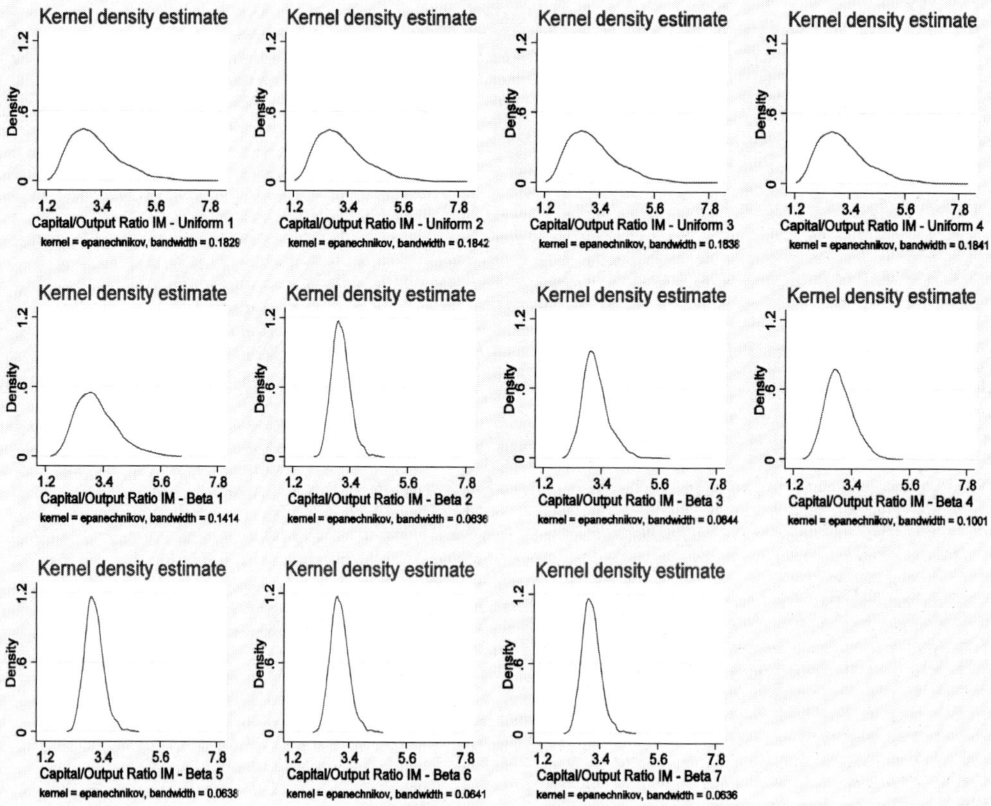

Figure 4. Capital/Output Ratios (IM) - Estimated Densities for the eleven Monte Carlo Experiments.

Interestingly, relaxing the borrowing constraint does not have any major impact on precautionary savings. This is shown in the first four panels of Figure (2). Even when agents can borrow a substantial amount of resources (up to three times their average income, as displayed in the related Figure 5), allowing them to better smooth consumption, this does not affect much precautionary savings: they do decrease, but not spectacularly so. At the same time, we can appreciate the effect of a highly left skewed prior: for the Beta(2,5) case reported in panel *B4* precautionary savings become a fraction of 1%. Income shocks that are not very persistent and with a very low variance, together with relatively impatient and not very risk averse individuals, explain the result.

9.4.2. Inequality and Exogenous Borrowing Constraints

It has generally been found that the Aiyagari 1994 model (and many of its variants) cannot generate high levels of wealth inequality, as described by the Gini index.[9] This result is confirmed by this global sensitivity analysis, with some caveats.

[9] In the interest of space, here I do not comment on the results for another inequality measure, the coefficient of variation. The estimated densities had the very same shape as the ones for the corresponding Gini index.

As for this model economy, Figure (6) shows that wealth tends to be more concentrated than in the calibrations reported by Aiyagari 1994. The first panel shows that, without access to borrowing, the Gini index does not vary much from its average value of 0.45, a value above the calibrations provided by Aiyagari 1994. In Table 4 we see that for the *U1* case the minimum is 0.24, the maximum is 0.55, and the s.d. is 0.03.

However, Aiyagari 1994 did not provide any robustness checks related to the borrowing limit. Panels 1 to 4 in Figure (6) present such an analysis. They consider the equilibrium effects of relying on a less strict exogenous borrowing constraint. When moving from the no borrowing case to a case where people can borrow, we can appreciate a first order effect on the Gini index. The average of the index increases, together with its dispersion, being equal to 0.56 in case *U4*. The minimum is now 0.25, the maximum is 0.88, and the s.d. is 0.09.

There is an important caveat related to this analysis. When allowing for negative assets, the Gini coefficient need not be between zero and one. In order to deal with a variable that can take negative values while giving a well-behaved Gini index which is comparable to the index for non-negative variables, I apply the adjustment proposed by Chen, Tsaur and Rhai 1982. Without correcting the index this way, the same panels would show an unpleasant feature: for some economies the Gini index would be above one.

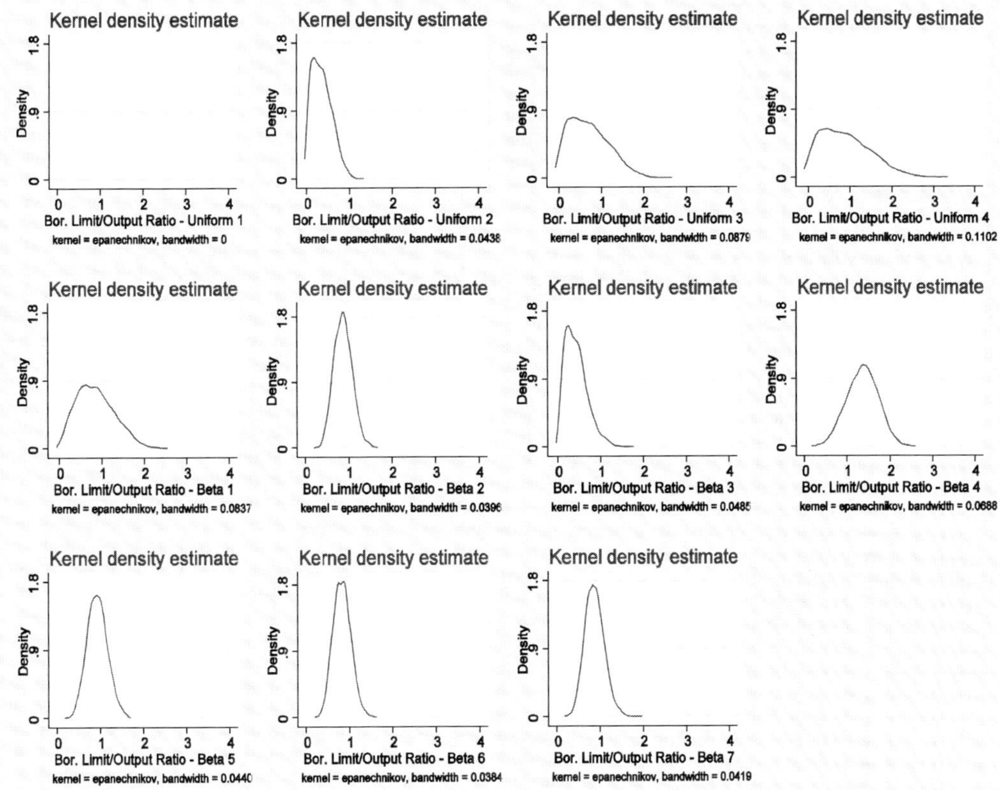

Figure 5. Borrowing Limit/Ouput Ratios - Estimated Densities for the eleven Monte Carlo Experiments.

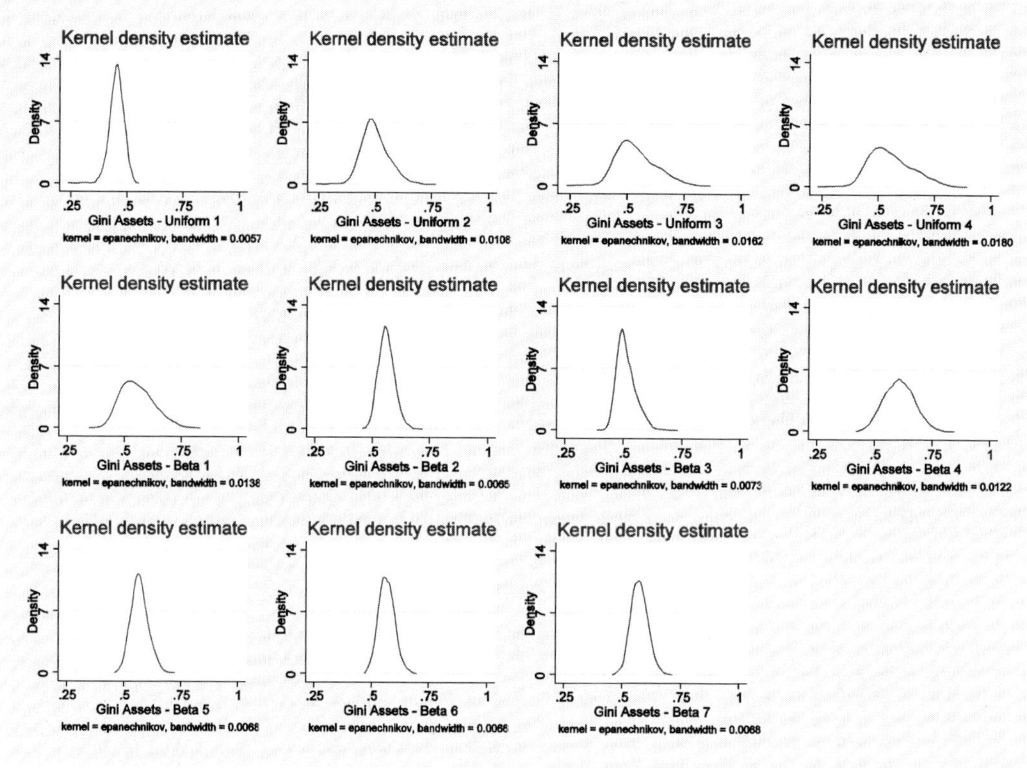

Figure 6. Wealth Gini Index - Estimated Densities for the eleven Monte Carlo Experiments.

Overall, the median wealth Gini index now ranges from 0.49 to 0.64, depending on the prior being considered. The first order effect of relaxing the borrowing constraint notwithstanding, these figures are still far below from what is observed in the data, a value of approximately 0.8. Hence, apart from a handful of calibrations in experiments *U3*, *U4*, *B1* and *B4*, the model cannot match the observed degree of wealth inequality in the U.S.

Figure (7) reports the percentage of households that have a negative net worth. As discussed above, the average (and median) values are consistent with what is observed in the data. Finally, the amounts of resources that can be borrowed compared to the average income appear to be plausible numbers, as the graphs in Figure (5) show.

9.4.3. (Global) Comparative Dynamics

With the results of the simulations at hand, I can perform a surrogate of comparative dynamics analysis. Rather than taking the derivative of the outcomes with respect to a parameter, I can run a regression, with either precautionary savings or the Gini index on the left hand side, and the corresponding simulated parameters on the right hand side.[10]

[10] This is a well posed exercise, because the parameters are changing randomly across replications. As a result, the source of variation is exogenous. However, when drawing the parameters, some spurious correlation is detected in the artificial dataset. I ran a 2SLS estimation procedure as well, and the results changed very little. All regressions include a constant term, not reported in the table.

Table 5. Comparative Dynamics - Linear Regressions, Standardized Beta Coefficients

	U1	*U2*	*U3*	*U4*	*B1*	*B2*	*B3*	*B4*	*B5*	*B6*	*B7*
Wealth Gini Index:											
α	-.54	-.59	-.58	-.57	-.57	-.60	-.49	-.60	-.58	-.60	-.59
β	-.21	-.20	-.17	-.16	-.16	-.16	-.15	-.18	-.17	-.14	-.14
δ	.48	.51	.49	.49	.50	.52	.42	.54	.50	.52	.51
σ	-.31	-.20	-.15	-.14	-.13	-.15	-.15	-.13	-.17	-.13	-.15
ρ_y	.34	.13	.06	0.04	0.07	0.12	0.19	0.01	0.11	0.07	0.08
σ_y	.30	.22	.19	.18	.15	.12	.06	.27	.14	.15	.15
b	-	-.43	-.54	-.56	-.55	-.56	-.68	-.42	-.55	-.53	-.53
R^2	*.83*	*.90*	*.91*	*.92*	*.94*	*.98*	*.94*	*.97*	*.99*	*.97*	*.97*
Precautionary Savings:											
α	-.16	-.13	-.11	-.10	-.09	-.09	-.10	-.07	-.08	-.08	-.05
β	-.15	-.14	-.14	-.14	-.12	-.13	-.18	-.07	-.11	-.09	-.08
δ	.01	-.02	-.04	-.05	-.05	-.07	-.04	-.04	-.06	-.06	-.06
σ	.48	.49	.50	.50	.49	.54	.45	.50	.52	.44	.42
ρ_y	-.21	-.21	-.20	-.20	-.15	-.13	-.48	-.01	-.13	-.51	-.50
σ_y	.65	.65	.64	.64	.73	.79	.63	.79	.79	.65	.66
b	-	.06	.10	.11	.09	.10	.10	.06	.10	.07	.07
R^2	*.74*	*.74*	*.74*	*.74*	*.83*	*.95*	*.88*	*.86*	*.95*	*.88*	*.89*

Table 5 shows the standardized beta coefficients of such regressions. It is interesting to notice that the same parameter has different quantitative effects on the two outcome variables. As for the wealth Gini index, the parameters that quantitatively have the largest impact on this inequality measure are the capital share α, the borrowing limit b, and the depreciation rate δ. The intuition goes as follows. The higher α, the higher the marginal product of capital and the higher the interest rate. This price change induces an income effect, which is going to kick in more likely for wealthier agents, driving the reduction in inequality. The depreciation rate has a similar, but opposite, mechanism with higher values of δ determining lower interest rates. The intuition for the borrowing limit is simple: the less resources people can borrow, the higher the precautionary savings and (possibly) the lower the range of wealth. This reduces the degree of wealth concentration both because the individuals save more, and because their maximum amount of debt is lower. Although their effect is quantitatively less important, a higher discount factor β and a higher risk aversion σ tend to compress inequality, by inducing all agents to save more. Surprisingly, the parameters driving the uncertainty in the economy, ρ_y, and σ_y, have a relatively low impact on the Gini index.

As for the precautionary savings, the parameters that affect them the most are the relative risk aversion σ and the s.d. of the income shocks σ_y.[17]

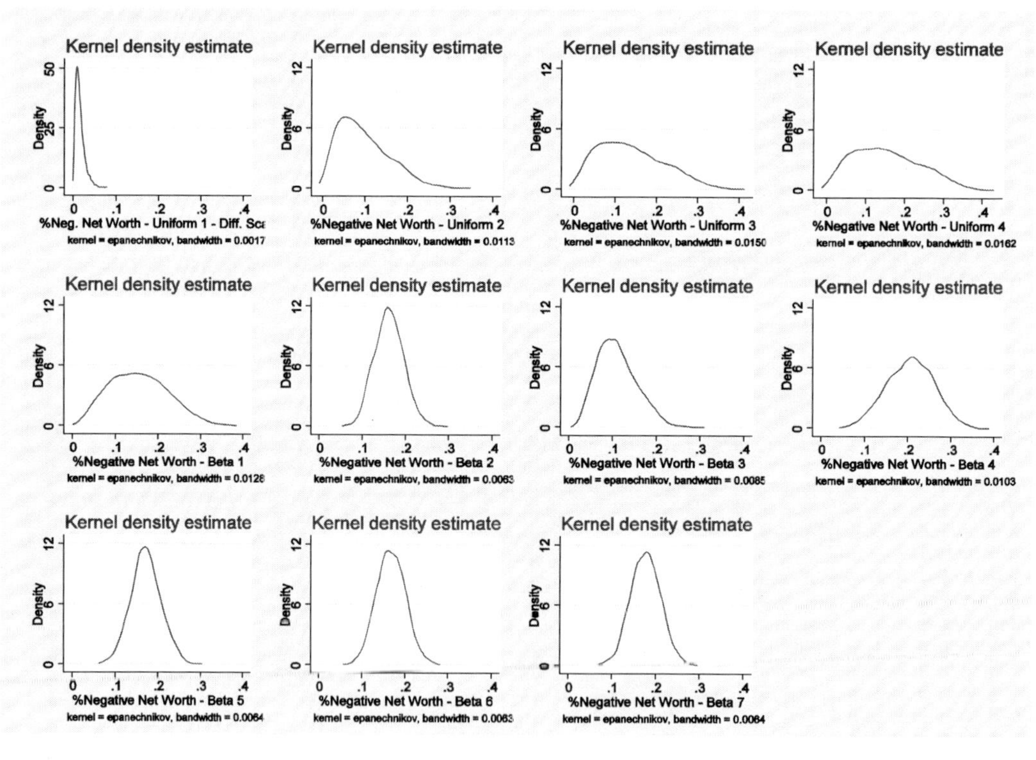

Figure 7. % with Negative Net Worth - Estimated Densities for the eleven Monte Carlo Experiments.

[17] A non-linear relationship is found for most parameters. The square of the parameters is significant in almost every experiment, but including them as additional regressors only marginally increases the explanatory power (usually by less than a percentage point). The point estimates of the square terms are positive only for α and σ in the Gini regressions, and for α, σ, σ_y and b in most of the precautionary savings ones. Higher order polynomials, as usual, introduce multi-collinearity for most specifications.

9.4.4. Robustness[2]

This subsection considers a robustness check for this global robustness analysis. With some abuse of language, it could be said that I am going to perform "robustness squared".

A possible source of criticism on the Monte Carlo calibration can come from the relatively naive way in which parameter vectors are selected, and their potentially counterfactual quantitative implications. Unfortunately, before solving the model, it is very complicated to guess which replications are going to "fail" in some relevant economic dimension. Restricting ex-ante the parameters' space and the set of plausible priors is somewhat challenging.

However, in order to tackle this conceivable issue, the simulations can be restricted ex-post. After having solved the model for 2,500 times relying on a specific prior, the simulations that imply implausible values for some simulated moments are dropped.

More precisely, the following restrictions on the capital/output ratio $\frac{K^{IM}}{Y^{IM}}$, the saving rate $\frac{\delta K^{IM}}{Y^{IM}}$, and the percentage of people in debt are imposed:

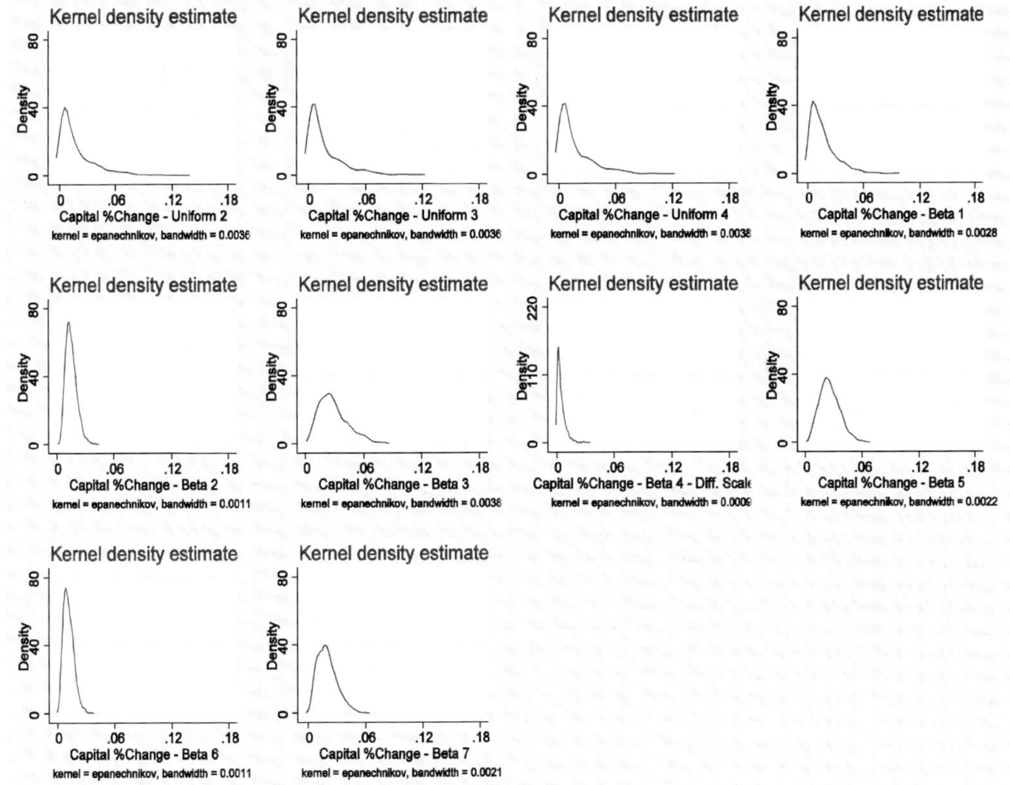

Figure 8. Precautionary Savings - Estimated Densities for the eleven *Restricted* Monte Carlo Experiments.

1) $1.5 \leq \frac{K^{IM}}{Y^{IM}} \leq 3.5$,

2) $0.125 \leq \frac{\delta K^{IM}}{Y^{IM}} = \frac{S^{IM}}{Y^{IM}} \leq 0.275$,

3) $0.05 \leq \text{Share of Households with a Negative Net Worth} \leq 0.2$.

Only the simulations that satisfy all three criteria at the same time are kept. This means that I am discarding between 595 replications for the case *B2* and 2,066 replications for the case *B3*. It goes without saying that, by default, the benchmark case *U1* does not satisfy the third requirement.

Tables 6 and 7 display the new parameters ranges, once the restrictions are imposed. With the uniform priors (Table 6), most parameters' bounds are not affected. The changes tend to be small, and affect mostly $\overline{\alpha}$, $\underline{\delta}$, and $\underline{b}$. Differently, with the Beta experiments (Table 7), most parameters' bounds do change. Some changes are relatively minor. For example, $\underline{\beta}$ is not affected much, its most stringent value now being 0.946 for case *B3*. Others are larger. For example, $\overline{\sigma}$ becomes 2.965 for case *B4*, $\underline{\sigma}$ becomes 1.162 for case *B3*, $\overline{\rho_y}$ becomes 0.957 for case *B2*, and $\overline{\sigma_y}$ becomes 0.190 for case *B4*.

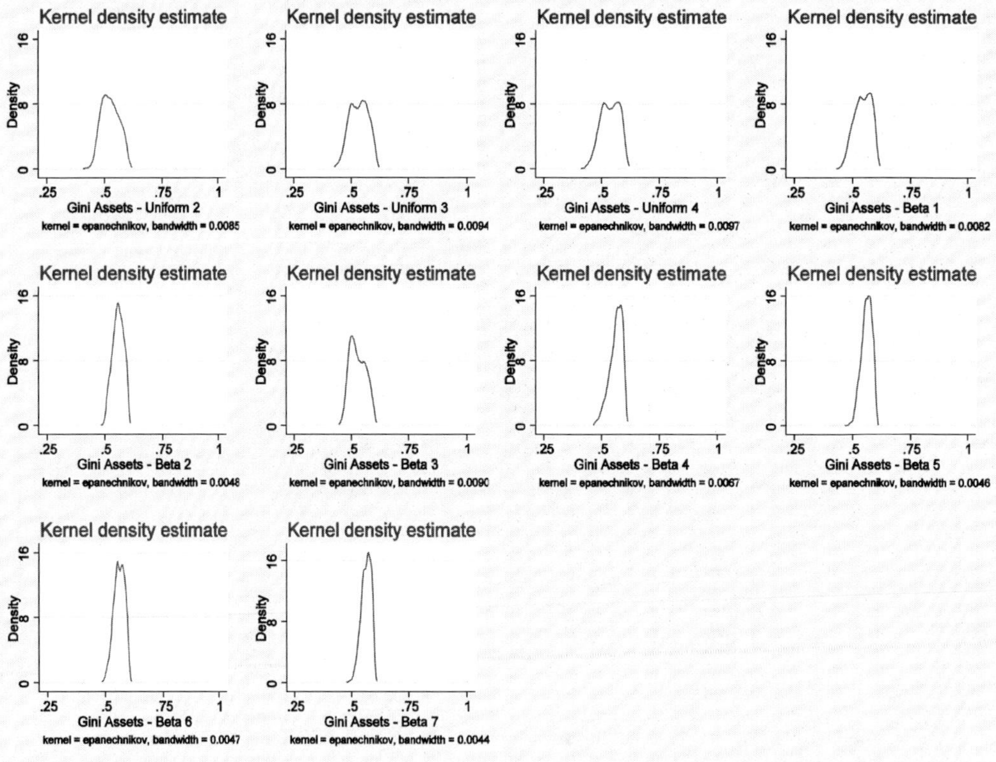

Figure 9. Wealth Gini Index - Estimated Densities for the eleven *Restricted* Monte Carlo Experiments.

Table 6. Model parameters and their support – Restricted simulations, Uniform experiments

Parameter	*U2*	*U3*	*U4*
$\overline{\alpha}$	0.445	0.445	0.445
$\underline{\alpha}$	0.250	0.250	0.250
$\overline{\beta}$	0.980	0.980	0.979
$\underline{\beta}$	0.940	0.940	0.940
$\overline{\delta}$	0.120	0.120	0.120
$\underline{\delta}$	0.039	0.039	0.039
$\overline{\sigma}$	3.499	3.499	3.499
$\underline{\sigma}$	0.501	0.501	0.501
$\overline{\rho_y}$	0.990	0.990	0.990
$\underline{\rho_y}$	0.800	0.800	0.800
$\overline{\sigma_y}$	0.250	0.250	0.250
$\underline{\sigma_y}$	0.040	0.040	0.040
$\overline{b}$	-0.01	-0.01	-0.01
$\underline{b}$	-1.00	-2.00	-2.50

Table 7. Model parameters and their support – Restricted simulations, Beta Experiments

Parameter	B1	B2	B3	B4	B5	B6	B7
$\overline{\alpha}$	0.441	0.408	0.417	0.407	0.408	0.408	0.408
$\underline{\alpha}$	0.260	0.287	0.280	0.254	0.291	0.287	0.287
$\overline{\beta}$	0.979	0.973	0.979	0.969	0.972	0.973	0.971
$\underline{\beta}$	0.941	0.946	0.946	0.940	0.946	0.946	0.946
$\overline{\delta}$	0.118	0.106	0.119	0.100	0.106	0.106	0.106
$\underline{\delta}$	0.038	0.051	0.054	0.037	0.051	0.051	0.051
$\overline{\sigma}$	3.477	2.968	3.478	2.965	2.978	2.968	2.968
$\underline{\sigma}$	0.602	0.992	1.162	0.505	0.992	0.992	0.992
$\overline{\rho_y}$	0.988	0.957	0.989	0.963	0.958	0.989	0.989
$\underline{\rho_y}$	0.801	0.832	0.841	0.800	0.832	0.836	0.836
$\overline{\sigma_y}$	0.247	0.207	0.248	0.190	0.248	0.204	0.248
$\underline{\sigma_y}$	0.040	0.078	0.079	0.042	0.080	0.078	0.080
$\overline{b}$	-0.09	-0.33	-0.04	-0.29	-0.33	-0.33	-0.33
$\underline{b}$	-2.40	-2.06	-1.60	-2.45	-1.95	-2.04	-2.00

Estimating the densities of the precautionary savings and the wealth Gini index only on the cases that meet the above criteria delivers Figures (8) and (9). From the graphs, we can conclude that the results presented in the previous sections are not altered substantially by these additional requirements.

Since precautionary savings were already found to be consistently small, Table 4 shows the results for the restricted cases only for wealth inequality. These are reported in parenthesis. As expected, the extreme values are eliminated by the restrictions: the minima increase by up to 19 points and the maxima decrease by up to 23 points. As a consequence, it is possible to conclude that all the replications that were found to get close to the data were failing in some of the three moments above. Finally, the median values of the wealth Gini index now range from 0.52 to 0.57, which is a small increase compared to the baseline case.

9.4.5. Discussion

At least four aspects of the analysis call for further discussion.

It goes without saying that also this empirical methodology is not free of potential issues. On the one hand, the priors' choice is somewhat arbitrary: there are no clear indications on how to specify these distributions. This aspect of the procedure can only be problem-dependent. Hence, in general, it is not possible to assess how much the results are going to be contingent on the specific priors that are being used. However, this obstacle can be partially overcome by considering several different cases, with relatively flexible and general specifications. At the same time, the priors should try to minimize the number of replications that entail implausible quantitative implications along some dimension. Finally, relevant empirical studies should be used as an additional guide in this choice. This was the approach attempted in this chapter.

Another complication is represented by the computational burden. The computational time that is needed to complete the procedure can become easily intractable, even with powerful computers. Unless a researcher has access to a supercomputer, the computational costs are still high: on a moderately fast desktop computer it takes approximately a week of continuous computer time to complete an experiment.[18] Hence, the solution method must strike a balance between the computational complexity, the implied numerical errors and its reliability. The last point seems to be quite important for non-linear models, such as the one we are considering here. The solution method has to guarantee that convergence problems are not going to riddle too many replications, especially for non-random subsets of the parameters. Otherwise, this could potentially bias the results. However, once a robust solution method is available, the extension for the Monte Carlo calibration methodology is extremely simple: it boils down to an additional outer loop where a vector of random parameters is drawn at each replication.

A natural question that arises is whether this method can be applied to models that are currently used in the quantitative macroeconomics field. The most recent contributions proposing *HA* models allow for several different sources of uncertainty and several layers of heterogeneity. It goes without saying that the more complex the model, the less likely the feasibility and the computational success of Canova's methodology. However, recent developments in numerical methods for economics can prove useful. The relatively simple and powerful endogenous grid method can cut the computational time substantially, as found by Barillas and Fernandez-Villaverde 2007, and Carroll 2006. The same objective can be achieved by more efficient simulation methods, as proposed by Nishimura and Stachurski

[18] Incidentally, this consideration made it unfeasible to rely on 10,000 replications, as in Canova's original contribution. However, for the *HA* economy under study, when considering only the first 1,000-1,500 cases, the results were very similar to the ones of the full experiment, suggesting that the number of replications is large enough.

2010. Trivially, a supercomputer with (say) 256 nodes would make the computational burden tractable for several richer models. However, this does not seem to be a reasonable requirement. Resource constraints and the added complexity of parallel computing in large clusters make this option not viable for many quantitative macroeconomists. A more plausible option for sizeable computational gains could come from GPU computing, as recently discussed in Aldrich, Fernandez-Villaverde, Gallant, and Rubio-Ramirez 2011, or by the family of Intel® Xeon Phi™ coprocessors that have recently been announced. These opportunities are yet to be carefully explored, but show great promise to meet the computational needs of quantitative macroeconomists.

Finally, unlike fully Bayesian methods (proposed for example by DeJong, Ingram, and Whiteman 1996), Canova's procedure does not allow the model to exploit the information contained in the data, to achieve a better parameterization while trying to reduce the model's parameter uncertainty. Unfortunately, for this class of models, the computation of the likelihood function (necessary to obtain the posterior distribution) still represents a daunting challenge, both in terms of data needs and computational complexity. Recently, a transition from pure calibration studies to analyses based on Bayesian principles has been taking place in simpler macroeconomic models with representative agents. In the coming years, it is plausible to expect that this method of empirical research will become very active also for equilibrium HA models, possibly with an approach along the lines of what is advocated by Geweke 2010.

CONCLUSION

This chapter contributed to the literature on calibration methods for equilibrium Heterogeneous-Agent macroeconomic models. It was shown that global (and more robust) sensitivity analyses are also feasible for such rich economies. Unless one is willing to rely on supercomputers, the computational costs are still high: on a moderately fast desktop computer it takes approximately a week of continuous computer time to complete an experiment. However, once a robust solution method is available, the extension for this simulation methodology is extremely simple: in essence, it boils down to an additional outer loop, where a vector of random parameters is drawn at each iteration.

It is worth stressing that this type of exercise is meant to complement more traditional calibration studies (or sophisticated structural estimation ones), not to substitute them. Once a quantitative theory is proposed and a set of empirically relevant values for the parameters are available, one can provide more robust evidence on the original quantitative findings, possibly suggesting in which cases some outcomes are more likely to be observed.

In a nutshell, the results of my experiments confirm Aiyagari 1994's findings: precautionary savings are small indeed, even when considering priors for the parameter uncertainty that are very different, and sometimes highly dispersed. Moreover, it was shown that varying the exogenous borrowing limit leads to relatively large changes in the wealth Gini index. However, even with increases in the median of the index of up to 15 points, the model is still very far from accounting for the wealth inequality observed in the U.S. economy. These findings are confirmed even when restricting the calibrations to satisfy some relevant features of the available macroeconomic data.

The parameters that quantitatively have the largest effects on determining the wealth Gini index are the capital share, the borrowing limit, and the depreciation rate. The parameters affecting precautionary savings the most are the risk aversion and the standard deviation of the income shocks.

A similar methodology can be implemented for other *HA* models, quantifying for example the welfare effects of eliminating the social security system, or the ones arising from changing the tax code. A positive aspect of this analysis is that, when applied to the evaluation of policy reforms, it provides distributions and ranges of welfare effects, an approach to empirical research pushed forward by Manski 1995, among others. I leave these extensions and modifications for future work.

APPENDIX A
THE MODEL AND ITS RECURSIVE REPRESENTATION

(1) Stationary Equilibrium

In this Section, I first define the problem of the agents in their recursive representation, then I define the problem of the firm, finally I provide a formal definition of the equilibrium concept used in this model, the recursive competitive equilibrium.

The individual state variables are the labor endowment $\varepsilon \in \mathcal{E} = [0, \bar{\varepsilon}]$, and asset holdings $a \in \mathcal{A} = [-b, \bar{a}]$. The stationary distribution is denoted by $\mu(\varepsilon, a)$.

(2) Problem of the Workers

The value function of an agent whose current asset holdings are equal to a, and whose current labor endowment is ε is denoted with $V(\varepsilon, a)$. The problem of these agents can be represented as follows:

$$V(\varepsilon, a) = \max_{c,a'} \{u(c) + \beta E_{\varepsilon'|\varepsilon} V(\varepsilon', a')\}, S.to$$

$$c + a' = (1 + r)a + w\varepsilon, a_0 \text{ given}, c \geq 0, a' > -b$$

$$\varepsilon' = \rho_y \varepsilon + \eta', \eta \sim \text{iid } N(0, \sigma_y^2)$$

Agents have to set optimally their consumption/savings plans. They enjoy utility from consumption, and face some uncertain events in the future. In the next period their labor income can go up or down, depending on the future realizations of the earnings shock η. Finally, they are subject to an exogenous borrowing constraint, $b \geq 0$.

(3) Problem of the Firm

The production side of the model is represented by a constant returns to scale technology of the Cobb-Douglas form, which relies on aggregate capital K and labor L to produce the final output Y.

$$Y = F(K, L) = K^{\alpha}L^{1-\alpha}$$

Capital depreciates at the exogenous rate δ and firms hire capital and labor every period from competitive markets. From the first order conditions of the firm we obtain the expression for the net real return to capital r and the wage rate per efficiency unit w:

$$= \alpha\left(\frac{L}{K}\right)^{1-\alpha} - \delta$$

$$w = 1 - \alpha\left(\frac{K}{L}\right)^{\alpha}$$

Notice that the marginal productivity of labor is always positive, hence firms will rely on the total sum of the efficiency units of labor. It follows that in the steady-state:

$$L = \int \varepsilon d\mu_{\varepsilon}$$

where μ_{ε} is the stationary distribution over the labor endowments implied by the markov process.

(4) Recursive Stationary Equilibrium

A recursive stationary equilibrium is a set of decision rules $\{c(\varepsilon, a), a'(\varepsilon, a), k = (K/L)\}$, value functions $V(\varepsilon, a)$, prices $\{r, w\}$ and stationary distributions $\mu(\varepsilon, a)$ such that:

- Given relative prices $\{r, w\}$, the individual policy functions $\{c(\varepsilon, a), a'(\varepsilon, a)\}$ solve the household problem and $V(\varepsilon, a)$ are the associated value functions.
- Given relative prices $\{r, w\}$, k solves the firm's problem.
- The asset market clears: $K = \int a d\mu(\varepsilon, a)$.
- The goods market clears: $F(K, L) = \int c(\varepsilon, a) d\mu(\varepsilon, a) + \delta K$.
- In equilibrium the measure of agents in each state is time invariant and consistent with individual decisions, and the stationary distributions $\mu(\varepsilon, a)$ satisfy:

$$\mu(\varepsilon', a') = \sum_{\varepsilon} \int_{a:a'(\varepsilon,a)=a'} \pi(\varepsilon', \varepsilon) d\mu(\varepsilon, a)$$

Appendix B - Computation

- All codes solving the economies were written in the FORTRAN 95 language, relying on the Intel Fortran Compiler, build 11.1.048 (with the IMSL library). They were compiled selecting the O3 option (maximize speed), and without automatic parallelization. They were run on a 64-bit PC platform, running Windows 7 Professional Edition, with an Intel Q6600 Quad Core 2.4 Ghz processor.

- On average, each replication takes 5 minutes. The actual computing time depends essentially on the discount factor β that is drawn, and on how far the initial guess on the interest rate is from the equilibrium one (from 1 to 13 iterations on the interest rate are needed to find each equilibrium). This means that, for any prior distribution, the whole Monte Carlo procedure takes between 8 and 13 days to complete. However, the Quad Core processor allows to run at least four cases simultaneously, with a 100% load, but without losing any computational speed. Obviously, more recent Quad and Hexa Core CPU's could handle even more simultaneous simulations, which would be completed in less time.
- In the actual solution of the model I need to discretize the two continuous state variables ε, a. As for ε, I rely on Tauchen's method, which approximates the AR(1) process for the efficiency units with a Markov chain. I use a seven-state approximation: this is a common number of points, as it strikes a good balance between approximation error and speed. As for a, I rely on an unevenly spaced grid, with the distance between two consecutive points increasing geometrically. In order to keep the computational burden manageable, I use 301 grid points on the asset space, the lowest value being the borrowing constraint and the highest one being a value $a_{max} > \bar{a}$ high enough for the saving functions to cut the 45 degree line ($a_{max} = 50$). This is done to allow for a high precision of the policy rules at low values of a, that is where the change in curvature is more pronounced.
- A collocation method is implemented. I look for the policy functions such that the residuals of the Euler equations are (close to) zero at the collocation points (which correspond to the asset grid). It follows that for all possible combinations of state variables I need to solve a non linear equation. A time iteration scheme is applied to get the policy functions, i.e. I compute the first order conditions with respect to a' and through the envelope condition I obtain a set of Euler equations, whose unknowns are the policy functions, $a'(\varepsilon, a)$. I start from a set of guesses, $a'(\varepsilon, a)_0$, and keep on iterating until a fixed point is reached (between 115 and 1,500 iterations are needed to reach a fixed point), i.e. until two successive iterations satisfy:
$\text{Sup}|a'(\varepsilon, a)_{n+1} - a'(\varepsilon, a)_n| < 10^{-8}, \forall\varepsilon$
- The stationary distributions $\mu(\varepsilon, a)$ are computed by simulating a large sample of 10,000 individuals for 3,000 periods, which ensure that the statistics of interest are stationary processes, and that they do not vary substantially when considering more individuals. For more details, see Rios-Rull 1999. This stage is particularly time consuming (approximately 30 seconds per iteration on the interest rate). Notice that in order to minimize the effect of sampling variability affecting the results, I do not rely on the simulation method used by Aiyagari 1994, the simulation of only one long history. For each statistic of the simulated sample, I consider the time average of their cross sectional values for the last 1,000 periods, rather than the value obtained from one long simulated history. As for the approximation method, I rely on a linear

approximation scheme for the saving and consumption functions, for values of the assets falling outside the grid.

- If the numerical procedure fails to converge in some of its steps, the related results are disregarded. This happens (rarely) when the time iteration procedure gets stuck in a cycle, rather than converging to the true policy functions.

APPENDIX C - MONTE CARLO ALGORITHM

The computational procedure used to solve the Monte Carlo experiments can be represented by the following algorithm:

1. Draw 2,500 combinations of parameters from their prior distributions and store them.
2. Calibrate the model by reading the first vector of simulated parameters and begin the model solution.
3. Generate a discrete grid over the asset space $[-b, \dots, a_{max}]$.
4. Generate a discrete grid over the efficiency units space $[\varepsilon_{min}, \dots, \varepsilon_{max}]$.
5. Get the aggregate labor supply L.
6. Guess on the interest rate r_0.
7. Get the capital demand k.
8. Get the wage rate per efficiency units w.
9. Get the saving functions $a'(\varepsilon, a)$.
10. Simulate the stationary distributions $\mu(\varepsilon, a)$.
11. Get the aggregate capital supply.
12. Check the asset market clearing; Get r_1.
13. Update $r_0' = \varpi r_0 + (1 - \varpi) r_1$ (with bisection).
14. Iterate until market clearing.
15. Get the consumption functions $c'(\varepsilon, a)$.
16. Check the final good market clearing.
17. Save the output and repeat from step 2. for all the 2,500 combinations of simulated parameters.

REFERENCES

Aiyagari, R., 1994. Uninsured idiosyncratic risk and aggregate saving. *Quart. J. of Econ.* 109, 659-684.

Aldrich, E., Fernandez-Villaverde, J., Gallant, R., Rubio-Ramirez, J., 2011. Tapping the supercomputer under your desk: Solving dynamic equilibrium models with graphics processors. *J. of Econ. Dyn. and Control* 35, 386-393.

Attanasio, O., Banks, J., Meghir, C., Weber, G., 1999. Humps and bumps in lifetime consumption. *J. of Bus. and Econ. Stat.* 17, 22-35.

Barillas, F., Fernandez-Villaverde, J., 2007. A generalization of the endogenous grid method. *J. of Econ. Dyn. and Control* 31, 2698-2712.

Budria Rodriguez, S., Diaz-Gimenez, J., Quadrini, V., Rios-Rull, J.V., 2002. Updated facts on the U.S. distributions of earnings, income, and wealth. *Fed. Res. Bank of Minneapolis Quart. Rev.* 26, 2-35.

Cagetti, M., De Nardi, M., 2008. Wealth inequality: data and models. *Macro. Dyn.* 12, 285-313.

Canova, F., 1994. Statistical Inference in Calibrated Models. *J. of Applied Econometrics* 9, S123-S144.

Carroll, C., 2006. The method of endogenous gridpoints for solving dynamic stochastic optimization problems. *Econ. Letters* 91, 312-320.

Chen, C., Tsaur, T., Rhai, T., 1982. The Gini coefficient and negative income. *Oxford Econ. Pap.* 34, 473-478.

DeJong, D., Ingram, B., Whiteman, C., 1996. A Bayesian approach to calibration. *J. of Bus. and Econ. Stat.* 14, 1-9.

Geweke, J., 2010. Complete and incomplete econometric models. *Princeton Uni. Press*, Princeton (NJ).

Guvenen, F., 2009. An empirical investigation of labor income processes. *Rev. of Econ. Dyn.* 12, 58-79.

Hansen, L. P., Heckman, J., 1996. The empirical foundations of calibration. *J. of Econ. Persp.* 10, 87-104.

Heathcote, J, Storesletten, K., Violante, G., 2009. Quantitative macroeconomics with heterogeneous households. *Ann. Rev. of Econ.* 1, 319-354.

Kydland, F., Prescott, E., 1982. Time to build and aggregate fluctuations. *Econometrica* 50, 1345-1370.

Kydland, F., Prescott, E., 1996. The computational experiment: an econometric tool. *J. of Econ. Persp.* 10, 69-85.

Manski, C., 1995. Identification problems in the social sciences, *Harvard Uni. Press*, Cambridge (MA).

Nishimura, K., Stachurski, J., 2010. Perfect simulation of stationary equilibria. *J. of Econ. Dyn. and Control* 34, 577-584.

Prescott, E., 1986. Theory ahead of business-cycle measurement. *Carn.-Roch. Conf. Ser. on Pub. Policy* 25, 11-44.

Rios-Rull, J.V., 1999. Computation of equilibria in heterogeneous agent model. In Computational methods for the study of dynamic economies: an introduction, (Marimon, R., Scott, A., Eds.), *Oxford Uni. Press*, Oxford (UK).

Quadrini, V., Rios-Rull, J.V., 1997. Understanding the U.S. distribution of wealth. *Fed. Res. Bank of Minneapolis Quart. Rev.* 21, 22-36.

Wolff, E., 1998. Recent trends in the size distribution of household wealth. *J. of Econ. Persp.* 12, 131-150.\

INDEX

T

U

V

W

Z